Science and Technology in a
Changing International Order

Also of Interest

† Available in hardcover and paperback.

Westview Special Studies in Social, Political, and Economic Development

Science and Technology in a Changing International Order:
The United Nations Conference on Science and Technology for Development
edited by Volker Rittberger

The authors of this volume review the issues involved in financing the development of endogenous scientific and technological capabilities in Third World countries and examine United Nations global conferences with regard to the options they offer for new international institution building. The authors also look at both contemporary patterns and future alternatives for Third World cooperation in science and technology for development and discuss the significance of the UN Conference on Science and Technology for Development (UNCSTD) for the advancement of women.

Dr. Rittberger is a professor of political science at the University of Tübingen (Federal Republic of Germany) and a Special Fellow at UNITAR (United Nations Institute for Training and Research). He is the author of numerous books and articles on international affairs.

Published for

UNITED NATIONS INSTITUTE
FOR TRAINING AND RESEARCH
(UNITAR)

INSTITUT DES NATIONS UNIES
POUR LA FORMATION ET LA RECHERCHE
(UNITAR)

Science and Technology in a Changing International Order

The United Nations Conference
on Science and Technology
for Development

edited by Volker Rittberger

Routledge
Taylor & Francis Group
LONDON AND NEW YORK

First published 1982 by Westview Press

Published 2019 by Routledge
52 Vanderbilt Avenue, New York, NY 10017
2 Park Square, Milton Park, Abingdon, Oxon OX14 4RN

Routledge is an imprint of the Taylor & Francis Group, an informa business

Library of Congress Cataloging in Publication Data
United Nations Conference on Science and Technology for Development (1979 : Vienna, Austria)
Science and technology in a changing international order.
(Westview special studies in social, political, and economic development)
"Published for the United Nations Institute for Training and Research."
Includes bibliographical references and index.
1. Underdeveloped areas—Science—Congresses.   2. Underdeveloped areas—Technology—Congresses.   3. Science—International cooperation—Congresses.   4. Technology—International cooperation—Congresses. 5. Science—Social aspects—Congresses.   6. Technology—Social aspects—Congresses.   I. Rittberger, Volker. II. United Nations Institute for Training and Research. III. Series.
Q127.2.U557   1979   338.9   81-11375
ISBN 0-86531-146-3   AACR2

ISBN 13: 978-0-367-28667-5 (hbk)

ISBN 13: 978-0-367-30213-9 (pbk)

Contents

vii

Preface

As part of its contribution to the 1979 United Nations Conference on Science and Technology for Development (UNCSTD) the United Nations Institute for Training and Research (UNITAR) organized an informal research group to assess various aspects of applying science and technology to development through the United Nations system. This research group was constituted in early 1978 at the initiative of Dr. Robert S. Jordan, former Director of Research at UNITAR, and was headed by Professor Volker Rittberger, a UNITAR Special Fellow. One of the activities of this group has been the production of a series of working papers on science and technology. These papers seek to provide preliminary analyses rather than definitive conclusions. Their purpose is to facilitate the access of others to the ongoing work of the group and to stimulate critical comments and reactions leading to further improvement of this work.

Science and technology for development has meant, for most of the industrially developed countries, the establishment of modern scientific institutions in developing countries, preferably through private channels. Many of the developing countries view science and technology for development as the abolition of all international barriers that hinder their access to the benefits of scientific-technological advance. However, the establishment of scientific institutions, massive transfers of technology, and the elimination of barriers to the flow of scientific-technological knowledge will not alone solve the problems facing the developing countries. The scientific and technological communities in the developing countries are small, and they are isolated from the mainstream of scientific inquiry. Furthermore, the educational systems of the developing countries in the areas of science and technology are insufficiently staffed, poorly equipped, and too small. Indeed, knowledge of science and technology and the benefits derived from it in both the industrially developed and the developing countries is limited to too few people, and the population as a whole is not affected by it.

Against this background, UNITAR is helping in the search for socially relevant methods of strengthening the technological autonomy of the developing countries and harnessing the potential of science and technology for the solution of global problems. This volume constitutes one such effort in this direction.

In keeping with UNITAR policy, the opinions expressed in this volume are the responsibility of the individual authors and do not necessarily reflect the views of the Institute, its Board of Trustees, or the organizations with which the authors are associated.

Grateful acknowledgement is to be given to Pamela D'Onofrio-Flores, whose substantive expertise and editorial work were indispensable to the UNCSTD project and to the subsequent preparation of this volume.

Davidson Nicol
Under-Secretary-General, United Nations
Executive Director, UNITAR

The Contributors

David W. Chu is Research Assistant for the United Nations Conference on Science and Technology for Development (UNCSTD) at the Council on International and Public Affairs and a doctoral candidate in political science at the Graduate School of Arts and Sciences, Columbia University, New York.

Pamela D'Onofrio-Flores is currently a Research Assistant to the Executive Director, The United Nations Institute for Training and Research (UNITAR). She has been a consultant on science and technology for development to the International Science and Technology Affairs Programme of the Council on International and Public Affairs (New York), the United Nations Office for Science and Technology, and The Research Policy Programme, The University of Lund (Sweden). In addition, Ms. D'Onofrio-Flores has provided expertise in the area of science and technology to such organizations as the United Nations Interim Fund for Science and Technology for Development and the Secretariat of the World Conference of the United Nations Decade for Women. She was a UNITAR Delegate to the 1979 United Nations Conference on Science and Technology for Development (UNCSTD) and has officially represented the Institute at such meetings as the United Nations Intergovernmental Committee on Science and Technology for Development and the United Nations Advisory Committee on Science and Technology for Development. Among her recent publications is *Science and Technology for Development: International Conflict or Co-operation — A Bibliography of Studies and Documents Related to the 1979 United Nations Conference on Science and Technology for Development*, (Research Policy Programme, University of Lund, Sweden).

Ward Morehouse is President of the Council on International and Public Affairs (New York) and Research Associate in the Southern

Asia Institute, Columbia University (New York). He was Visiting Professor at the Administrative Staff College of India in Hyderabad (1969-1970) and the Research Policy Institute, University of Lund in Sweden (1976-1977). He has been a Consultant for the Technology Transfer Division, United Nations Conference on Trade and Development (1977-1978) and is currently a member of a Research Team on Science and Technology for Development, the United Nations Institute for Training and Research (UNITAR). His major and recent publications include: *Science in India: Institution Building and the Organizational System for Research and Development* (Bombay: Popular Prakashan, 1971); *Science, Technology and National Development in India: The Uses, Non-Uses and Misuses of Research* (Madras: Dr. Sir A. L. Mudaliar Endowment Lectures in Technology, University of Madras, 1972); and *A New Science and Technology Order* (New Brunswick: Transaction Books, 1979). He is also the editor of *American Labor in a Changing World Economy* (New York: Praeger Publishers in co-operation with the Carnegie Endowment for International Peace, 1978).

Dr. Davidson Nicol is currently an Under-Secretary-General of the United Nations and the Executive Director of the United Nations Institute for Training and Research (UNITAR). Before assuming these positions, Dr. Nicol was Vice Chancellor and Head of the University of Sierra Leone in Freetown, a Danforth Fellowship lecturer in African Affairs at the Association of American Colleges, the Permanent Representative of Sierra Leone to the United Nations, the High Commissioner of Sierra Leone to the United Kingdom, and the Ambassador of Sierra Leone to Sweden, Norway and Denmark. He gained first class honours and two doctorates in Natural Science and Medicine from the University of Cambridge, United Kingdom, and was elected a Fellow of Christ's College. Dr. Nicol has published numerous articles and several books on scientific, literary and political subjects. He is the author of *Africa: A Subjective View* (Longman's, 1964) and the co-editor of *The United Nations and Decision-Making: The Role of Women* (UNITAR, 1978). With L. F. Smith at Cambridge, England, in 1959 he elucidated the structure of human insulin.

Volker Rittberger, Ph.D., is currently a Special Fellow at UNITAR and a professor of political science at the University of Tübingen, Federal Republic of Germany. He studied law and political science at the Universities of Freiburg i. Br. and Geneva as well as at Stanford University. From 1971 to 1973 he was a senior researcher at the Frankfurt Peace Research Institute. He has taught political science at the Universities of Freiburg i. Br. and Heidelberg. His research interests cover German foreign policy, disarmament, development, and international organization and politics. His English-language publications include *Evolution and International Organization* and *The Foreign Policy of West Germany* as well as contributions to scientific journals and collective works. Professor Rittberger represented UNITAR at the sessions of the Preparatory Committee for UNCSTD and was a member of the UNITAR delegation to UNCSTD (Vienna, 1979).

Francisco R. Sagasti, Ph.D., M.Sc., also holds two engineering degrees from the National University of Lima, Peru. He served as Vice-Chairman of the Board of the Industrial Technology Institute in Lima (1972–1977) and worked at the same time as co-ordinator of an international research project on technological development in Third World countries under the auspices of the International Development Research Centre in 1973 and involving the participation of ten countries and more than 150 researchers. He is the author of many articles and several books on science and technology. His latest book, *Technology, Planning and Self-Reliant Development: A Latin American View*, has recently been published by Praeger Publishers, New York. He was also a Delegate to the 1979 United Nations Conference on Science and Technology for Development. At present, he is Associate Director of GRADE, an independent research centre, and Managing Director of TEAD, a consulting firm in Lima, Peru. He also teaches at the Universidad del Pacífico and is a member of the Board of the Peruvian National Research Council. Dr. Sagasti was awarded The 1980 United Nations Peace Medal and The Paul Hoffman Award for outstanding contributions to development.

Abbreviations

ACAST	Advisory Committee on the Application of Science and Technology
CSTD	Committee on Science and Technology for Development
ECOSOC	Economic and Social Council
FAO	Food and Agriculture Organization
GDP	Gross domestic product
GNP	Gross national product
IDRC	International Development Research Center (Canada)
ITINTEC	Industrial Technology Institute (Peru)
NIC	Newly industrializing country
NIEO	New International Economic Order
OECD	Organisation for Economic Co-operation and Development
OPEC	Organization of Petroleum Exporting Countries
R&D	Research and development
S&T	Science and technology
SAREC	Swedish Agency for Research Cooperation with Developing Countries
TCDC	Technical Cooperation among Developing Countries
TNC	Transnational corporation
UNCSTD	United Nations Conference on Science and Technology For Development
UNCTAD	United Nations Conference on Trade and Development
UNDP	United Nations Development Programme
UNESCO	United Nations Educational, Scientific and Cultural Organization
UNIDO	United Nations Industrial Organization
UNITAR	United Nations Institute for Training and Research

Introduction

Davidson Nicol
Volker Rittberger

The United Nations Institute for Training and Research (UNITAR) was established to give the United Nations a tool to reflect upon problems of a global or universal nature, which frequently can only be solved by international collaboration, often with the participation of United Nations agencies or bodies. It is therefore natural that UNITAR has had a long history of interest and involvement in studying the impact of scientific and technological advances on world affairs in general and the access of developing countries to technological innovations in particular.

During the past decade, UNITAR commissioned and published a series of research reports detailing certain of the problems of and prospects for the transfer of technology from developed to developing countries both on an industry-specific and on a country-specific basis. Special attention was given to key industries, such as the pharmaceuticals, petrochemicals, semiconductors, paper and pulp, and automotive industries.

As a parallel to these studies on the transfer of technology from developed to developing countries, UNITAR also sponsored the most extensive empirical investigation to date into the motivations of and factors influencing those who form part of the "brain drain" from developing to industrially developed countries. These studies have had a seminal influence on the evolving international discussion about the transfer of technology and the brain drain and the problems that they pose for the development of developing countries.

Other topics relating to basic concerns of the United Nations Strategy for the Second Development Decade and to its Declaration and Programme of Action on the Establishment of a New International Economic Order have figured prominently among the policy-oriented activities of UNITAR at both research and training levels.

1

Thus, the results of global conferences, changing aspects of international law, and the possibilities of and obstacles to creating a new international order have been among the subjects of UNITAR publications and seminars over the last five years.

The emphasis on science and technology, including aspects relevant to energy resources and their development, has been reflected in some major conferences that UNITAR has recently organized jointly with other bodies. A conference on "Petroleum and Gas Energy" at Laxenburg, Austria, in 1976 was followed by a conference on "Microbial Energy Conversion" in Göttingen in the Federal Republic of Germany. In 1978 a conference on "Small Scale Mining" was held in Mexico and in 1979 an important conference on "Heavy Crude Oil and Tar Sands" was held in Edmonton, Alberta, Canada. In December of 1979 another large conference on "Alternate Sources of Energy" took place in Montreal, Quebec, Canada. These UNITAR conferences, which have involved thousands of delegates from the industrially developed countries and from the developing countries at both the decision-making and the technical levels, have demonstrated that practical cooperation between the North and the South is possible and has already started in United Nations bodies, including, in these particular cases, UNITAR.

When the United Nations General Assembly decided in 1976 to convene a special world conference on science and technology for development before the end of the 1970s, UNITAR immediately developed an active interest in the preparations. In accordance with its mandate as the preeminent reflective body within the United Nations, the Institute participated, within the available means, in a wide variety of preparatory activities for the United Nations Conference on Science and Technology for Development.

In addition to providing expert input at a number of meetings, the UNITAR contribution to the preparatory work for the Conference consisted, above all, in establishing a series of Science and Technology Working Papers and also in organizing a seminar on "Financial Arrangements for the Promotion of Science and Technology for Development," which took place July 1979 at Schloss Hernstein in Austria.

The Working Papers were written by a group of scholars and experts from developed and developing countries in their individual capacities; some appear in this volume. The themes run the whole gamut of issues that were before UNCSTD. However, the papers selectively accentuate some aspects on which there has been relatively little policy-oriented research to date. Chapters 1 and 2 address the basic issue of the scientific and technological dependence of developing countries and analyze ways of overcoming this condition; Chapter 3 focuses specifically on the crucial question of strengthening cooperation among developing countries as a way of increasing their scientific and technological self-reliance; Chapter 4 discusses a range of options for financial and institutional arrangements that may facilitate the implementation of the Vienna Programme of Action adopted by UNCSTD; Chapter 5 deals with the role of women in science, technology, and development; and Chapter 6 is devoted to an analysis of how the preparatory process of the Conference contributed to the articulation, among various social groups, of interests in and needs for the development and use of science and technology in a social and political context. The interest of the Federal Republic of Germany in these issues is evidenced by its major support for this extensive research.

The Working Papers confirm the fundamental working assumption of the Conference, namely that there exists a general, yet also graduated, backwardness of developing countries in the field of science and technology that should no longer be tolerated and that this has external as well as internal sources that are mutually reinforcing. Thus, whereas the developing countries themselves must mobilize their own resources to achieve a fair measure of self-reliance, the international community, and the industrially developed countries in particular, have an interest in and the capability of assisting the developing countries. The industrially developed countries can also assist the competent and relevant international institutions in their efforts to overcome extreme disparities in the distribution of scientific and technological capacities and in the distribution of the benefits that accrue from scientific-technological development.

There is an emerging danger of a tacit and certainly unintended alliance between those in the developing countries (and elsewhere) advocating the dissociation of developing countries from cooperation with the industrially developed countries and those in the developed countries arguing for an approach of "benign neglect" toward the developing countries. The conflict potential inherent in such a constellation is great. It is therefore an urgent obligation of enlightened leadership in both North and South, East and West, to fashion a workable political programme that meets the urgent needs of developing countries in the field of science and technology and that will ultimately benefit all countries.

The Working Papers give considerable attention to the politics of strengthening and utilizing scientific and technological capabilities. They generally agree that there is an interactive or interdependent relationship between the process of strengthening capacities and the process of setting the social goals and priorities that these capacities are supposed to serve. It is obvious that strengthened scientific and technological capabilities can be misused for power and prestige purposes and also for increasing social inequities.

The necessity of facing squarely the question of social goals and priorities is not meant to divert attention from the task of building, adapting, strengthening, and efficiently managing scientific and technological potential and capabilities in developing countries. Indeed, UNITAR has placed great emphasis on studying a very basic aspect of capacity building, namely appropriate financial arrangements for the follow-up activities outlined in the Vienna Programme of Action.

Our professional work on this subject has made it apparent that a basic consensus is emerging that developing countries need, in addition to mobilizing and utilizing more effectively their domestic resources, an increased inflow of financial resources from the industrially developed countries, both to strengthen their indigenous scientific and technological capabilities and to derive maximum benefit from scientific and technological innovations achieved elsewhere. To be of real value to long-term development planning in developing countries, this transfer of resources should occur in an assured, continuous, and predictable manner. As investments in the field of science and technology very often have a long maturity period and, moreover, are not, in many cases, of a self-liquidating nature, public funds of considerable volume will be required, in addition to private funds, to stimulate the growth of science and technology in developing countries and to assure its momentum in the industrially developed countries.

This volume is intended to further elucidate the problems and prospects, options and alternatives for the rational use of science and technology in the public interest and for the national, regional, and international machinery required toward that end.

1
Science and Technology as an International Order and Development Issue: An Overview

Volker Rittberger

1. The Quest for New International Orders

Ever since the proclamation of the New International Economic Order (NIEO) by the United Nations General Assembly in 1974[1] the notion of new international orders has crept into all attempts at restructuring the existing patterns of international relations in a wide variety of issue areas. The NIEO has occupied center stage for some time and has stood (*pars pro toto*) for the basic interests of Third World nationalist leadership groups in transforming the "old order" of First World dominance in world affairs. Also, a series of demands for other sectorally defined new international orders has been formulated in various multilateral diplomatic fora.

In addition to the NIEO, another major initiative, under the aegis of the United Nations, to achieve a sectorally defined new international order is represented by the work of the Third United Nations Conference on the Law of the Sea (UNCLOS III) toward a reshaping of the international maritime order.[2] This attempt at sectoral reform with a global scope constitutes a most interesting example of the difference between the "old" and a "new" order. The "old order" of maritime affairs was based primarily on the liberal principle of the "freedom of the high seas," which, in fact, has favoured those who were capable of effectively exercising this freedom. Instead, the envisioned "new order" will have to strike a balance among competing principles—"freedom of the high seas," ocean space as the "common heritage of mankind," national sovereignty over coastal and adjacent waters—and will therefore have to rely on political-administrative mechanisms of collective regulation, intervention, and control to mediate between widely differing users' interests in ocean space.

A third prominent battleground that has emerged, so far, in the North-South confrontation over a new international order is related to information and communication.[3] Within both the United Nations and the United Nations Educational, Scientific and Cultural Organization (UNESCO), spokesmen for the developing countries have criticized the existing international patterns of producing, distributing, and circulating information as biased in favour of the North, whose information agencies and communications media tend to dominate international and national information markets, including those of the Third World. Accordingly, the developing countries seek to initiate a series of internationally agreed-upon measures to reduce the inequities in this field.

Reviewing these trends, one may formulate the proposition that the quest for new international orders represents, above all, a challenge to the liberal concept of international relations that has been characteristic of the post–World War II world under U.S. hegemony. Typically, the protagonists of this concept have been blind to the discrepancy between the theoretical premises that are thought to legitimize a social order based on market relations and the realities of international stratification. The proponents of the old liberal order have for too long paid insufficient attention to the blatant disparities of resources and capabilities among nations that have kept many of them, to various degrees, in a state of dependency on one or more crucial dimensions. Thus, the insistence on the unfettered operation of market forces as the regulating principle of peaceful intercourse among nations may be viewed, under these circumstances, as being not very far from cynicism.

However, the formulation of this proposition needs to be accompanied by the observation that it is by no means clear what the ultimate, and even the intermediate, outcomes of this challenge to the old liberal order are likely to be. As to the short run, it appears that the foremost – intended or unintended – consequences would be a strengthening of state control over social processes, particularly those extending beyond national boundaries. Such a tendency would be ambiguous in sociopolitical terms insofar as the nature of the political regime determines whether increased state control over social processes amounts to a higher level of popular self-government and mass welfare or to a reinforcing of elite and/or bureaucratic power. Whatever the result may be in a given case, it is fairly obvious that, in most developing countries, there exists a relatively broad consensus about the need to strengthen state power as a necessary prerequisite of defense against dependence-generating or -maintaining forms of interaction with more developed countries. Put differently, it seems important to keep in mind the close interrelationship between the demand for transforming the old liberal order and the interest in strengthening state power in the quest of the developing countries for new international orders. It goes without saying that the interest in strengthening state power is predicated on implementing a strategy of enhancing indigenous capabilities to assess, select, acquire, and produce goods and services that are instrumental for development. Chief among them are scientific and technological capabilities. Yet, the direction and quality of development will be determined by the uses to which these capabilities are put.

2. On the Uses of Science and Technology for Development

"Science and technology"[4] has become, for many, a magic formula in today's world: Whenever social problems (in the broadest sense) arise, the call for more scientific research and technological know-how is one of the most prominent responses on the part of contemporary ruling elites in many parts of the world.[5] Certainly, it is not the only option available to ruling elites who are confronted with social problems for which established political-administrative routines do not provide solutions that command broad compliance by those affected. In some instances, ruling elites may try simply to ignore the problems; in others, they may tend to silence those who attempt to analyze and raise public awareness of the problems; and in still others, they are committed to coping with the problems through policy planning (of one variety or another). Currently, one prominent, if not dominant, approach to social problem solving through policy planning is the technocratic one.[6] Typically, technocratic "problem-solvers," when faced with difficult political choices, resort to an analysis that almost invariably points up the complexity of the situation, which, in turn, justifies, indeed necessitates, more research as a precondition for successfully grappling with the problems in question. Expert committees are then set up, conferences and symposia are convened, new research and development (R&D) programmes are initiated, and the role of science and technology as a productive force of prime importance is once again emphasized. For ruling elites who adopt the technocratic approach, a readiness to promote science and technology as problem-solving tools serves to fulfill multiple functions: First, it diffuses the problems, at least in the short run; second, it furthers the emergence of a socially potent clientele among the intelligentsia while placating those interest groups that might feel hurt by a direct attack on the problems; and third, it is an act of self-legitimization, demonstrating that they are on top of the situation and responsive to the exigencies of their times.

There is no question that science and technology can help solve social problems and that they have done so in the past. Yet the bias toward scientism, i.e., the belief that there is a science-based technical solution to every social problem, that is inherent in the technocratic approach to social problem solving should be scrutinized for its usually unstated political implications: The prominence attributed to science and technology by technocratic elites tends to obscure the possibility that it may be the very technocratic approach that prevents or impedes the solution of urgent social problems, because the acceptance of certain solutions of other than a purely technical nature might entail consequences that amount to "rocking the boat," i.e., transforming the social status quo. Moreover, R&D efforts may be directed, as a result of public policy, toward false problems and/or toward the wrong solutions. This often preempts opportunities for independent critical analysis and for public consideration of options the implementation of which would imply far-reaching changes in the socioeconomic and political structures of a given country. The sociopolitical selectivity that is built into the technocratic ideology of scientism may entail a volume and a distribution of social costs that cannot be expected to be balanced automatically by the volume and the distribution of benefits accruing from promoting scientific and technological innovation systems both at the national and the international levels.

Take, for example, the case of hunger and malnutrition.[7] Undoubtedly, this constitutes a crucial social problem in many parts of the world. All relevant international statistics, incomplete and inaccurate as they may be in various details, agree that starvation continues to plague a very large proportion of people in developing countries and that there is little prospect for improvement over the next few years. The most prominent solution to which the technocratic problem-solvers (including those from developing countries) have turned, at least until very recently, is the call for more R&D in agriculture (and, of course, for an expansion of food aid). What happened, and still happens, is that more R&D in agriculture may lead to increased productivity without necessarily decreasing hunger and malnutrition. For increases in agricultural productivity help eradicate hunger and malnutrition only if the increased output is channeled toward satisfying domestic needs and if the needy can acquire the means to share in this increased output, i.e., if the discrepancy between real needs and effective demand is being significantly reduced. In fact, the problem of hunger and malnutrition cannot be attributed necessarily to lagging agricultural productivity and, therefore, to insufficient R&D inputs only, as the technocratic problem solvers like to see it. As Jacques Chonchol aptly put it:

la faim ne disparaîtra pas si l'on s'en tient à des politiques qui exagèrent l'importance de la production, de l'investissement et de l'aide.

Même quand la production s'accroît, l'offre alimentaire pour les pauvres peut decroître en raison d'une orientation excessive vers l'exportation, la production pour satisfaire les goûts importés et onéreux des minorités privilégiées, les pertes après récolte, et un système de distribution insuffisant.[8]

The origin and perpetuation of hunger and malnutrition can be located in a highly stratified and disequilibrated social structure rather than through a productivist explanation.[9] Yet, such a sociopolitical structure also guarantees the technocratic elites and their allies a hegemonic status. As a result, the serious consideration of agrarian reform and nonexploitative rural development has usually been given very low priority; and it has been only very recently that the issue of hunger and malnutrition has been explicitly linked to that of agrarian reform in an international forum with indisputable political legitimacy.[10] The reasons for this neglect in the past are obvious in the case of big landowners—both the old landed elite and the transnational corporations entering agribusiness—and in the case of ruling groups in developing countries, who often rely on the former, to varying degrees, as an important support group. Moreover, large segments of the relevant scientific and technological communities in industrially developed countries form part of this coalition working against a serious consideration of agrarian reform. This is because their expertise is geared primarily toward large-scale farming, which is more capital- and, therefore, more technology-intensive than small-scale farming. However, this is not to assert that the scientific and technological communities cannot make a constructive contribution to the development of the agricultural sector in developing countries, i.e., to help the peasant and small farmer to improve food production to satisfy domestic needs. Indeed, the improvement of small-scale farming may not be feasible without substantial R&D efforts either. Yet, the R&D required here would be of a different kind, such as upgrading traditional village technologies, developing the techniques of eco-farming, designing self-reliant cooperative institutions, and the like.[11]

In summary, when entering a discourse about "science and technology for development" it should be made clear that development is not just more (usually measured in aggregate terms) of certain goods and services. To be useful as a category of social science and praxis the concept of development requires a specification in terms of "who gets what," etc., as a result of marshaling scientific-technological resources for solving critical social problems.

3. Science and Technology in United Nations Concepts of Development

Science and technology play a great role in documents (resolutions, declarations, programmes of action, etc.) of the United Nations system calling for a reshaping of the international system with a view toward improving the well-being of developing countries and of their poorer segments (people and regions) in particular. The first major document with this special emphasis on science and technology as levers for accelerating the development process of developing countries has been the International Development Strategy of the General Assembly for the Second United Nations Development Decade.[12] Some of the central points in this document are:

- the expansion of the capability of developing countries to apply science and technology to development and to reduce the technological gap (paragraph 60);
- the increase, by developing countries, of their expenditure on research and development averaging 0.5 percent of their GNP per annum (paragraph 61);
- the development of appropriate technologies for developing countries (paragraphs 61–63);
- the strengthening of international cooperation and assistance to promote science and technology in developing countries (paragraphs 62 and 63);
- the establishment of a programme for promoting the transfer of technology to developing countries (paragraph 64).

Shortly after the adoption of the Action Programme for the Second United Nations Development Decade the United Nations Advisory Committee on the Application of Science and Technology to Development (ACAST) published its World Plan of Action for the Application of Science and Technology to Development,[13] which was later complemented by regional plans of action. These documents were meant to give substance to the general formulations of the strategy. Even though it was a masterpiece of analytic precision and technical expertise, the World Plan had little practical impact: Most developing countries simply lacked both the necessary indigenous scientific and technological capabilities and an appropriate sociopolitical system for implementing it.

A renewed emphasis on science and technology for development as an important issue area within the context of North-South relations was found in the four principal resolutions of the United Nations General Assembly ushering in, at the programmatic level, the era of the New International Economic Order.[14] The Declaration and the Programme of Action on the Establishment of a New International Economic Order expressed support for the principles of "giving to the developing countries access to the achievements of modern science and technology" and of "promoting the transfer of technology and the creation of indigenous technology for the benefit of the developing countries in forms and in accordance with procedures which are suited to their economies" (Declaration, paragraph 4(p); see also Programme of Action, section IV). Elaborating on these principles, the Charter of Economic Rights and Duties of States formulates in Article 9: "All States have the responsibility to co-operate in the . . . scientific and technological fields for the promotion of economic and social progress throughout the world, especially that of the developing countries." And in Article 13 the charter provides that "Every State has the right to benefit from the advances and developments in science and technology for the acceleration of its economic and social development" (paragraph 1). It is noteworthy that the Charter stresses in Article 13, paragraph 3, the necessity of international cooperation for strengthening the scientific and technological infrastructures of developing countries. Finally, the resolution of the Seventh Special Session on Development and International Economic Co-operation devotes a long section (section IV) to the issue area of science and technology for development containing, in a nutshell, the agenda and programme of work of the recommended United Nations Conference on Science and Technology for Development (paragraph 7), as well as references to the need for negotiations about a code of conduct for the transfer of technology and about revising the international regime of industrial property rights.

Aside from these resolutions, virtually all major global conferences organized by the United Nations and/or its specialized agencies have referred, in their final documents, to science and technology as important factors in the development process. In a study on such conferences during the 1970s, John Logsdon and Mary Allen discovered the following trends pointing to areas of consensus about the nature and function of science and technology in the development process:[15]

- an emphasis on creating the appropriate balance, for each Member State, between the development of indigenous scientific and technological capabilities and access to the scientific and technological resources of other countries;

- a growing interest in technology described as "practical," "appropriate," or "intermediate" as a particular focus for United Nations activity;

- a constant concern with the migration of trained scientific and technical personnel away from their native countries;

- an emphasis on the roles developed countries should play on various scientific and technological issues by:

  1. devoting more attention in developed country research and development programmes to problems of concern to developing countries;

  2. providing access for developing countries on equitable terms to the scientific and technological resources of developed countries;

  3. providing greater assistance, both on a bilateral basis and through multilateral institutions, to the developing countries in their attempts to develop indigenous scientific and technological capabilities;

- increasing emphasis on the importance of technical co-operation among developing countries;

- particular emphasis on research, training and information requirements related to achieving basic human needs.

From this brief summary of United Nations statements on science and technology for development, there emerge the contours of a concept of development that takes a view of science and technology that dispenses with the examination of the social prerequisites, conditions, and consequences of relying on science and technology for solving development problems. Put differently, one cannot escape the conclusion that there is a strong technocratic ring in these United Nations statements. Whether or not the outcome of the United Nations Conference on Science and Technology for Development (UNCSTD)[16] has departed from this direction and, if so, to what extent, will be discussed in Chapter 6. Suffice it to indicate here that a more complex problem perception seems to pervade the documents adopted at this Conference.

4. Alternative Perspectives on the Role of Science and Technology in the Development Process

The role of science and technology in the development process can be conceptualized, in summary fashion, in terms of three general perspectives: optimism, pessimism, and realism. One might ask which perspective approximates most closely the prevailing view in United Nations concepts of development.[17]

It is suggested that the perspective characterized as scientific-technological optimism captures best the way in which United Nations documents perceive the relationship between science and technology, on the one hand, and the development process, on the other. It is also the lifeblood of the technocratic approach. This means that they proceed from the assumption that science and technology have played a crucial and, in general, beneficial role in the development of today's economically advanced countries and regions. Thus, they take the view that the fruits of applying science and technology to development, hitherto enjoyed by a few countries only, should be made available to all countries and regions of the world. To this end they propose a two-pronged approach to strengthen the scientific-technological capacities of developing countries and to transfer technology to them as a means of fostering, above all, economic growth.

Quite obviously, the United Nations documents referred to above and the development strategy conceived by them have little, if anything, in common with the pessimist's perspective on the role of science and technology in the development process. This pessimistic view derives from the observation "that, while there is more technology in our lives than ever, major social problems do not seem to be any more tractable than before. On the contrary, it seems that the main problems of modern industrialized societies have developed because of modern technology and not in spite of it. . . ."[18] It is interesting to note in this context that, in developing countries, this view appears to have less support among the national elites than in the developed countries. Rather, "even where there is scepticism in the Third World about Western societies, enthusiasm for Western science and technology nevertheless seems to prevail."[19] In summary, then, if scientific-technological pessimism regards scientific-technological change as a, or even the, source of many critical social problems rather than as a means to solve them, then this perspective stands in clear contradiction to the prevailing United Nations concept of the role of science and technology in the development process.

One might still wonder, however, whether this juxtaposition of the optimistic and pessimistic perspectives is really helpful in critically assessing the concepts of the role of science and technology in the development process. It seems plausible to make the following assumptions forming the basis for another—the realist—perspective:

- that science and technology play a pervasive role in social dynamics;

- that science and technology can help solve social problems and contribute to development;

- that the application of science and technology to the solving of social problems frequently entails unintended consequences that may be detrimental to development efforts; this occurs most often when the social implications of scientific and technological change have been neglected;

- and that science and technology can be, and have been, used for wasteful and destructive purposes.

The notion of the dialectical nature of scientific-technological change inherent in the realist view refers to the historical simultaneity of constructive and destructive qualities in human labour and interaction. In the words of a recent publication by the World Council of Churches:

[Knowledge] is power to create and destroy, power loaded with promises and threats. . . . Science through its contribution to understanding liberates people from many forms of ignorance and superstition. Technology liberates them from many physical constraints and insecurities. . . . Yet in the face of all these promises, science and technology appear to many people as threats. . . . It is not only destructive human purposes that turn science and technology from promise to threat. Even well-intended uses of technology have unintended consequences that perplex or frustrate the people who initiated them.[20]

Denis Goulet, comparing technology to a "two-edged sword," emphasized two elements of technology's ambiguous impact on society: "(a) modern technology is simultaneously the bearer and the destroyer of precious human values, and (b) although it brings new freedom from old constraints imposed by nature, tradition, or ancient social patterns, technology also introduces new determinisms into the life of its adepts."[21] The critical aspect of this evolutionary process today is, of course, not the wholesale rejection of scientific and technological innovation; rather it is the need for a theoretically informed and ethically grounded social praxis that successfully prevents the destructive qualities of scientific-technological change from overpowering the constructive ones. One attempt to lay the ground for such a praxis is represented by the World Council of Churches' favoured three concepts of justice, participation, and sustainability.[22] Somewhat similarly, the Scheveningen Report on a New International Development Strategy singled out participation and decentralization, demilitarization, respect for cultural alternatives, and sustainability as principles of a social praxis oriented toward a nonexploitative development path.[23]

Admittedly, these ideas have been alien to the United Nations documents on the International Development Strategy and a New International Economic Order that were cited previously. The optimistic ring pervading all of them proved largely incompatible with a readiness to consider the externalities and social costs of scientific and technological change. Yet, this blindness has not been total. In certain respects — the arms race may be taken as the most prominent example — the question of wasteful and destructive uses of science and technology has been dealt with in United Nations documents in a very explicit and critical (albeit sometimes rhetorical) way. A case in point is the concern about the misapplication of the world's scientific and technological resources in that a large proportion of them (both financial and human) are absorbed by military R&D work.[24] Yet, this cannot detract from the more general observation that United Nations documents dealing with issues of social and economic development have not projected a coherent, realistic picture of the nature of scientific-technological change and its role in society. It should be noted, however, that designing such a picture would presuppose that a generally accepted definition of the core problems in the issue area of science and technology for development exists. The opposite is the case: The United Nations is exposed to the cross-pressures of competing problem perceptions to which different actors subscribe in different ways.

5. Competing Conceptualizations of the Main Issues in Policymaking for Science and Technology for Development

The problems relating to the generation, distribution, and use of science and technology for development, both at the national and at the international levels, can be conceptualized in a number of ways. The most prominent and by no means mutually exclusive perspectives on how to approach these problems can be described as follows.

A. Dependencia-*Approach*[25]

The core problem is seen localized in the wide gap that exists between the scientific and technological capabilities of the industrially developed countries and those of the developing countries and that tends to be reinforced by the pursuit of a liberal "open door" policy toward the importation of foreign technologies. The relative, if not absolute, backwardness and dependence of developing countries in applying science and technology to their development is indicated by the low level of supply of indigenous scientific and technological capabilities and, equally important, by an insufficient domestic demand for their products. Importation of foreign technologies, in whatever form, and emigration of skilled manpower is therefore widespread in Third World countries, frequently exacerbated by sociopolitical conditions that pose formidable obstacles to, or even cause a regression in, the growth of indigenous scientific and technological capabilities. Given this lack of capabilities, developing countries are often not even in a position to select and absorb the most suitable technologies available to them from foreign sources.

The solution to the problem as stated is expected, above all, from the "technological transformation of the Third World,"[26] particularly the endogenization of science and technology innovation systems. To achieve this, a policy mix is considered necessary that combines national actions of developing and industrially developed countries, specifically and differentially designed to strengthen the indigenous scientific and technological capabilities of developing countries by the establishment of international regulatory regimes for the transfer of technology and foreign investment, as well as by the setting up of new mechanisms for the transfer of resources to help finance the accelerated growth of national and local science and technology innovation systems in the Third World. Implicit in this approach is the view that the actual welfare deficits of the developing countries, which find their expression in hunger, diseases, illiteracy, unemployment, etc., will be eliminated, or at least markedly reduced, in due course.

B. Global Problems Approach[27]

Here, the emphasis in defining the core problem shifts from the disparities and discrepancies in scientific and technological capabilities between North and South to the general inadequacy of existing science and technology potential to cope with global problems. "Global problems" have been defined as the consequences of "the increasing stress we are placing on both the life-supporting biophysical eco-systems within which human life exists and the dynamic imbalances that are developing between human activities and the socio-economic-political environment within which these activities are carried on."[28] To give a more concrete image of what is meant by "global problems" the following distinctions have been offered:

- "Global manifestations": "the mounting world demand for food, materials, energy, waste management capacity, and social services; growing shortages of such resources as easily recoverable minerals, fresh water, favourable soils and suitable climate; attending high rates of inflation, unemployment and underemployment, increasing armament industries and nuclear proliferation; and more general signs of social disarticulation. . .;"

- "local manifestations" in industrially developed and more advanced developing countries: "pollution and contamination of the various media (air, water, soil, etc.) and associated health effects, and excessive as well as wasteful exploitation of renewable and nonrenewable resources";

- "local manifestations" in the poorer developing countries: "soil erosion and depletion, especially of marginal lands, massive deforestation in the tropical and semi-tropical areas, desertification in the semi-arid zone, the reappearance of diseases (such as malaria) that had been contained and even eliminated for many years, and the near-epidemic eruption of others (schistosomiasis, for example)."[29]

Another distinctive characteristic of "global problems" is their "international ramifications," i.e., "they cannot be solved independently from actions by other countries."[30] The response to this problem identification tends to be one demanding that, in general, more resources be devoted to research and development programmes and institutions and that a leading role be given to the scientific and technological communities in shaping policies that address themselves to "global problems." Some adherents of this approach realize, however, "that the role of science and technology in coping with global problems is not unambiguous": Science and technology are both part of the problems and part of the solutions.[31] Moreover, this approach is not oblivious to the highly skewed distribution of world scientific and technological potential. Rather, the differences between industrially developed and developing countries are recognized as being part of the overall problematique. Therefore, the strengthening of the science and technology base through national as well as international measures is strongly supported, on the one hand, and the full utilization of the existing capacities of industrially developed countries is urged, on the other. In any case, the solutions offered by this approach focus on the common predicament of humanity, the notion of "spaceship earth," rather than on the differences of access, by people or nations, to the goods and of exposure to the ills of the contemporary world.

C. Social Control Approach[32]

This identification of the core problem departs from the previous ones in that it addresses itself explicitly to the social uses and abuses of science and technology as well as to their differential impacts both within and among nations. In the words of the 1975 Dag Hammarskjöld Report: "Development of science and technology has become primarily a political and social issue, not a technical one. Producing technology, in the present international structure, means producing instruments of control and influence over other individuals, firms and nations. The capacity of technology to transform the nature, orientation and purpose of development is such that *the question of who controls technology is central to who controls development.* . . ."[33]

Underlying this approach is the perception that the generation, distribution, and application of science and technology do not occur in a social vacuum. If left to market forces, to the often self-regarding interests of the scientific and technological communities, or to the power strategies of bureaucracies, scientific and technological advancement may not lead to the betterment of the human condition in general and of the quality of life of the disadvantaged in particular. Instead, scientific and technological change will tend to reinforce existing inequities, enhance the wasteful utilization (and sometimes an equally wasteful nonutilization) of resources, and contribute to the destruction of humanity's living conditions.

The preferred solution is not the mobilization of a populistic antiscience and antitechnology movement that seems to haunt those who subscribe to scientific and technological optimism. In the final analysis, this approach would seem to suggest the *social-democratic* endogenization of science and technology rather than the merely *national* endogenization, the solution the *dependencia* approach is suggesting. Quite obviously, such a solution is more easily advocated than accomplished, if only in embryonic form. However, the merit of the social control approach is that, at the very least, it highlights the need for a political regime for science and technology, both at the national and at the international levels, that is capable of giving effective representation to diverse, even conflicting, interests concerning the collective uses of science and technology in developing as well as industrially developed countries.

If we turn to international political fora encompassing both developing and industrially developed countries on a formally equal footing, it is not surprising to discover that the *dependencia* approach tends to pervade their deliberations and resolutions. As most political leadership groups in Third World countries subscribe to development nationalism in one form or another, the *dependencia* approach to the analysis of underdevelopment and to the formulation of development strategies stressing the exogenous causes and conditions of underdevelopment represents to them a "natural" and potent ally at the ideological level. This is not to contend that there are no significant segments within Third World elites that do not share the view—which is more widely held in the scientific and technological communities and political leadership groups of the industrially developed countries, however—that national and international science and technology policymaking should also be guided by considerations germane to the global problems approach. On the other hand, the social control approach has yet to find a significant number of champions in international political fora. Yet, this fact is not too difficult to understand when one accepts the view of these fora as largely bureaucratic interfaces that tend to protect the formal or informal prerogatives of the participating bureaucracies as well as those of one of their most important partners in policymaking, i.e., the established intelligentsia and its transnational networks.[34]

6. Science and Technology Disparities Among Nations and Their Consequences

A. *The R&D Gap Between North and South*

The wide recognition that the *dependencia* approach to defining the main issues in international science and technology policy has found among actors in international political fora derives from the unquestionable fact that, with regard to the multidimensional development gap between North and South, it is the scientific and technological dimension that gives rise to grave concern. Even though we do not yet have a fully satisfactory data base for determining exactly the distribution of the world's R&D resources, it is nevertheless possible to sketch a global picture without distorting the main features and trends. Jan Annerstedt, who has done pioneering work in the compilation and analysis of data on R&D resources on a worldwide basis, posed the problem as follows: "Today, is the majority of the countries in the world forming a research desert, and can the remaining countries be seen as a small number of R&D oases?"[35] In answering this question he has based his analysis of the international division of labour in science and technology on two indicators: the distribution of world R&D expenditures and of world R&D manpower.[36] His general assessment runs as follows:

> At present the concentration of R&D to a small number of countries is one of the major features of global inequality. Less than 3 per cent of the world's R&D expenditures and just a little more than 11 per cent of its R&D scientists and engineers were controlled by the developing countries in 1973. Six nations (USA, USSR, Japan, the Federal Republic of Germany, France, United Kingdom) employed 72 per cent of the world's researchers and spent 83 per cent of R&D funds. The USA and USSR alone accounted for about 60 per cent of the global R&D expenditures.[37]

If one looks at Annerstedt's data more closely, several stark features stand out. First, the developing countries taken as a whole

are indeed far behind in the mobilization of R&D resources. There
has been some progress in reducing the gap between industrially
developed and developing countries over the last two decades; yet
the extent of this change has been comparatively small and can be
attributed to a very small number of countries in the Third World.
The share of developing countries in the world's R&D resources appears to be more sizable when based on the distribution of manpower rather than on the distribution of expenditures. However,
this should not be too surprising given the salary differentials as
well as the differences in investment per scientist or engineer between industrially developed and developing countries. At the same
time, the higher share of R&D manpower provides a glimpse of the
important potential that the developing countries already possess,
even if it is not yet fully utilized. The growth of this potential
becomes even more impressive when one considers the trends in
higher education: The share of the developing countries in the
number of students enrolled in higher education rose from 17 percent of the world's total in 1950 to 25 percent in 1972.[38]

A second feature commanding attention is the indication of considerable variations in scientific and technological capabilities
among both industrially developed and developing countries.[39]
Within the group of industrially developed countries, it is
noticeable that the socialist countries, which account for 33 percent
of the world's R&D expenditures and for 42 percent of the world's
R&D manpower, devote a relatively larger proportion of their gross
national product (GNP) — more than 4 percent — to R&D activities
than the other industrially developed countries: In Western Europe
and North America the equivalent figures are 1.56 percent and 2.35
percent, respectively. As regards R&D manpower, the share of the
socialist countries is significantly higher than the expenditures
would suggest whereas Western Europe's share is somewhat lower
(21.4 percent of world expenditures and 14.5 percent of world manpower) and the share of North America (Canada and the United
States) even more so (33.7 percent of world expenditures and 20.2
percent of world manpower). Comparing these three groupings of
industrially developed countries, it is also found that the number of
scientists and engineers per economically active person in Western
Europe is less than half that in North America and only one-third
that in the socialist countries.

If we look at the group of developing countries, it becomes apparent that the Latin American countries tend to spend three times
as much per economically active person on R&D activities as

African or Asian nations. Also, Arab and South and South-East Asian countries commit a share of their GNP to R&D activities that is lower than the average of all developing countries. If we look at the R&D manpower data, the stratification pattern within the Third World does not change much. Again, the Latin American countries come out on top, closely followed by the Arab countries, whereas the countries of South and South-East Asia and those of sub-Saharan Africa lag far, if not very far, behind.

Moreover, even within Third World regions scientific and technological potential is very unevenly distributed. As a rule, a very small number of countries control most of the already limited R&D resources of the Third World altogether. In the Latin American and Caribbean region, Brazil is the leading nation in R&D activities, followed by Argentina and Mexico. In the Arab world, the leading role is played by Egypt; and in South and South-East Asia, India claims more than half of all R&D expenditures and manpower of this region. In Black Africa, however, the level of R&D activities is so low that it would not be meaningful to search for a comparable pattern of stratification.

In any event, in analyzing these variations between and among both industrially developed and developing countries, the point of departure, i.e., the dramatic gap in scientific and technological capabilities between North and South, should not be lost sight of. It is the developing countries that rely heavily, if not exclusively, on externally created knowledge and skills feeding, more or less uncontrolled by them, into their educational systems and productive sectors.

B. *The Burden of Scientific and Technological Backwardness*

The scientific and technological backwardness of developing countries, which has been documented by the data on the international distribution of R&D resources, places a heavy burden on them: While attempting to benefit from the scientific and technological achievements of the more advanced countries through transfer processes they have to endure undesirable consequences that transfer processes between more advanced countries would be unlikely to entail. According to Frances Stewart,[40] four sets of these undesirable consequences can be distinguished, as follows.

(1) Transfer Cost. The total cost of payments for the transfer of technology from industrially developed to developing countries has not been calculated so far, one reason being that such an endeavour

would face methodological and practical obstacles that are difficult to overcome. Among these obstacles one may note, for example, the difficulty of sorting out the return on technology in payments for imported capital goods or qualified foreign personnel and in profit remittances. The figures of transfer cost that *are* available stem from a study of the UNCTAD secretariat that, therefore, concentrates on the direct costs of overt technology transfer.[41]

These direct costs, which usually are foreign exchange costs and thus put a strain on the balance of payments, amounted to about US $1,500 million in 1968. This sum was equal to

- 0.5 percent of gross domestic product (GDP),
- 5 percent of exports (excluding petroleum),
- 37 percent of public debt service payments,
- 56 percent of annual inflow of direct private foreign investment, and
- 250 percent of public R&D expenditures in developing countries.

As to their likely increase, the UNCTAD secretariat estimated that by 1980 the direct costs would reach US $9,000 million.

Considering the indirect costs of technology transfer, such as "the payments for technology embodied in the cost of imported equipment and intermediate goods and in remitted profits,"[42] the best data are informed guesses. Taking her point of departure from the developing countries' expenditures on imports of machinery, equipment, and chemicals and assuming that 10 percent of these expenditures was a return on the technology embodied in these products, Frances Stewart held that by including the indirect costs the total costs would be more than doubled.[43]

The heavy financial burden of the developing countries as a result of technology transfer is directly related to their condition of technological dependence: "In the first place, it is technological dependence that leads to the necessity for the net import of technology and the consequent net payment. Secondly, the situation of technological dependence is responsible for the very weak bargaining position of many developing countries *vis-à-vis* technology suppliers, and consequently for the poor terms exacted."[44]

(2) Loss of Control. Technologically backward countries are particularly vulnerable to imperfect market conditions that are typical of international technology markets: They not only have to pay

more for the technology that they buy abroad, they also have less control over the nature and use of the imported technology in enterprises operating on their territories. An extreme case is the totally foreign-owned firm for which all major decisions on investment, operations, marketing, etc., are taken at headquarters far away. However, the nationalization of the ownership of assets does not suffice to reduce or avoid technological dependence. On the contrary, even firms that are owned and managed by nationals of a given developing country, but that rely on foreign technology, find "that the content of the technology agreements are such that most of the power of independent decision-making is taken out of the hands of the local owners and managers."[45] These restrictive practices in technology transfer agreements have come under increasing attack in international political fora: The demand of the developing countries for negotiating an internationally binding code of conduct on the transfer of technology reflects their desire to repatriate at least part of the decision-making power that they signed away in their bilateral dealings with powerful technology suppliers.[46]

(3) Unsuitable Technology. As most imported technology was originally designed for use in industrially developed countries, its transfer to developing countries may render it less suitable there because conditions and needs are likely to differ significantly.[47] This amounts to a truism. However, the fact that there has been a lengthy debate, occasionally charged with strong emotions, about "appropriate technology"[48] suggests that this issue affects fundamental interests and aspirations. Emphasizing the suitability or appropriateness of foreign technology implies, to be sure, an inclination to place restrictions on technology imports.

It should not be surprising that the domestic and foreign interests associated with the "modern" sectors of industry and agriculture in developing countries are bound to be opposed to such a concept. On the other hand, the political appeal of the concept of suitability or appropriateness of technology derives its validity from the wide variations in the level of economic and technological development among the Third World countries (and even within them). Although this concept does not per se exclude the option of the importation of an advanced technology even in a least developed country, it points to both the necessity and the difficulty of making a considered choice among available foreign technologies for a given economic and/or social purpose.

(4) Inhibition of Local Innovation. Scientifically and technologically backward countries are caught in a vicious circle: Backward-

ness is both a result and a cause of the lack of indigenous scientific and technological capabilities. And without such indigenous capabilities the local capacity to innovate in the productive sectors is likely to be negligible. This inhibition of local innovation is related to two phenomena typical of most developing countries. One is the missing opportunities for "learning-by-doing" in the experimental development and practical application of new technologies: ". . . if developing countries are to build up the human skills and the institutional systems that are needed to reduce the extent of technological dependence, firms, laboratories and engineering organisations must have opportunities for learning-by-doing. There has to be some way of coping with the short-run costs of inefficiency and inexperience in their activities. . . . "[49] The other widespread phenomenon is the mutual estrangement between local scientific and technological institutions and local firms: " . . . there are repeatedly cases where local scientific and engineering laboratories in developing countries have been able to develop a technology to the point of commercial production – only to find that local firms prefer to license a precisely similar technique from abroad, in spite of higher financial costs."[50]

These burdens of scientific and technological backwardness and dependence represent a fundamental challenge to the formulation of any international development strategy for science and technology. It is to them that such a strategy must give priority attention and for which it must seek measures of alleviation.

7. Goals and Problems of an International Development Strategy for Science and Technology

It follows from the logic of even a minimalist concept of development, i.e., that a widening of the gap between North and South on a number of relevant dimensions must be halted and the trend reversed, that the debates in international political fora, such as UNC-STD, have centered, so far, on the definition of goals, methods, and techniques of altering the conditions of scientific and technological backwardness and dependence in the Third World. This emphasis must be seen and understood in the light of the paucity of scientific and technological capabilities in most developing countries. Discussions about "global problems" and the social control of science and technology tend to be perceived as meaningless by those who lack the elementary prerequisites for effective participation in problem-solving and policymaking processes in the field of science and technology. However, a purely capacity-oriented approach toward an international development strategy for science and technology would not be satisfactory either, because every strategy decides, at least implicitly, "who gets what? etc.," whether or not these questions are explicitly raised and answered. Thus, an unbiased strategic analysis will not separate growth from distribution or vice versa but look at them as a set of interrelated variables.[51]

A. Targets for Building Indigenous Scientific and Technological Capabilities

Building indigenous scientific and technological capabilities in developing countries is a process of mobilizing dormant or underutilized domestic (human and material) resources and of drawing on external assistance for facilitating the mobilization of these resources. However, although external assistance may be said to be necessary in the abstract, its concrete manifestations may be less than beneficial, if not outright detrimental. Targets have been set for external financial assistance by industrially developed countries for the building of indigenous scientific and technological capabilities of developing countries, yet these targets have not been met.[52] Thus, if developing countries had trusted these targets they would have founded their planning on false promises and have had to pay for the additional costs of waste and frustration. Moreover, even if targets had been met, how would one have protected the building of indigenous capabilities from inappropriate donors' influences? The multilateralization of external assistance does not necessarily enhance the autonomy of the recipient country because assistance is still given for certain project categories that are externally conceived and may, or may not, fit the specific development needs of a developing country.[53]

It might be useful, therefore, to consider, at least, a pooling of external assistance funds for science and technology and a mechanism that would entitle a developing country to financial assistance after having explained and justified, under a kind of cross-examination by peers, its financial needs for implementing one or more core elements of its sectoral development programme for science and technology.

This individualized programming approach toward setting targets for building indigenous scientific and technological capabilities as well as toward allocating external assistance takes into account the wide variations among developing countries noted previously. Building indigenous capabilities might mean, in one case, to begin with the creation of a viable scientific and technological infrastructure. This would imply, above all, the establishment of large-scale and differentiated education and training facilities and programmes for technicians, engineers, and scientists, the initiation of bottom-up institution-building, and, last but not least, the fostering of R&D work oriented toward satisfying local users' needs. Yet, the success of these measures would depend on whether or not they contributed to a synthesis between traditional knowledge and skills, on the one hand, and modern science and technology, on the other. In another case, promoting indigenous scientific and technological capabilities could mean the adaptation and reorientation of the existing, relatively advanced scientific and technological institutions and capabilities of a developing country. Often, the scientific and technological establishments in more advanced developing countries possess fairly extensive links with the relevant international scientific and technological communities and make significant contributions to the advancement of theoretical knowledge. However, frequently they remain separated in their work and aspirations from their national societies. There are several reasons for this, one, perhaps the most important, being the lack of demand pressure by local enterprises. Thus, it would be necessary both to reshape the national systems of higher learning and academic research to bring them into closer contact with domestic agricultural and industrial production and to direct potential local users of R&D work toward potential domestic suppliers. The social and political obstacles that a government pursuing these policies would encounter should not be underestimated, however.[54]

It is obvious that such a qualitative operationalization of targets for building indigenous scientific and technological capabilities would tend to provoke controversies about the substance of development goals to be served by these capabilities. Conversely, purely quantitative target setting facilitates the avoidance of substantive conflicts because it is compatible with different qualitative policy objectives and allows a spurious consensus between actors holding qualitatively different views about desirable goals and processes of development. However, experience on the basis of the International Development Strategy of 1970[55] and the Lima Declaration of 1975[56] suggests that "such targets may be misleading, inadequate for some of the countries they are directed to or beyond the reach for others: in short, ineffective."[57] Thus, one may conclude that the developmental relevance of promoting indigenous scientific and technological capabilities hinges on the socioeconomic and political direction of this process. Within developing countries the setting of qualitative targets has to be organized as an interactive process between government and support groups for building indigenous capabilities. It is through such an interaction that the socioeconomic and political contents of building indigenous capabilities will be elaborated and that this process will take firm root in a broadly based web of domestic interests committed to self-reliant development. This does not necessarily presuppose a system of government of the type that exists, for example, in most countries of the Organisation for Economic Cooperation and Development (OECD). However, institutions and processes that ensure adequate representation of national interest groups, already organized or yet to be organized, vis-à-vis the central and regional administrations in developing countries may be an elementary ingredient in self-reliant development, particularly in the field of science and technology.

As to the aid relationship between industrially developed and developing countries, the insistence on giving priority to qualitative target setting may involve the risk of forgoing a sizable part of the external assistance that otherwise could have been expected, because substantive disagreements between donor and recipient countries would then become more frequent. Even though most industrially developed countries will express their readiness to contribute to the strengthening of scientific and technological capabilities in developing countries, many of them extend their assistance on the assumption, if not the condition, that it is likely to suit their foreign economic interests and their own scientific and technological development planning. Thus, to avoid inappropriate external influences over the building of their indigenous capabilities, leadership groups in developing countries will have to realize that they have a stake in widening the possibilities for coalition formation, across the North-South divide, among reform-minded regimes, movements, parties, and so on, to facilitate greater compatibility between the perceived interests of donor and the needs of recipient countries.[58]

B. Technology Transfer and Its Dilemmas

Complementary to building up indigenous scientific and technological capabilities and bridging the time lag, another method advocated in United Nations documents[59] to overcome technological backwardness and dependence calls for a rapid and massive increase in the transfer of technology from industrially developed to developing countries on commercial or, preferably, concessional terms.[60] Even though certainly representing a plausible approach to counteracting the effects of exogenous factors that have generated or maintain technological backwardness and dependence in the Third World, transfer of technology is, on all accounts, beset with ambivalences and loaded with seemingly inescapable contradictions.

One position holds that, in general, transfer of technology will reproduce, at best, the existing dependency relationship between industrialized and developing countries while increasing the differentiation among, and the structural heterogeneity within, the latter. According to Johan Galtung, "The total picture . . . is one of transfer of technology as structural and cultural invasion, an invasion possibly more insidious than colonialism and neo-colonialism, because such an invasion is not always accompanied by a physical Western presence. . . ."[61] Dieter Ernst identified a relationship between growing transfer of technology and "a new industrialization scenario which, superficially, may fulfil some of the expectations prevailing, for instance, in some recent declarations of the 'Group of 77,' (but) may in fact turn out to fulfil nearly all the preconditions to significantly increase the technological dependence of these countries." As a result, Ernst predicted "a qualitative intensification of global patterns of technological dominance and dependence."[62] Following this approach, advocating an increase in the transfer of technology from industrialized to developing countries may well imply, objectively, an acquiescence in the continued technological superiority of the industrially developed countries. Moreover, it is to be expected that the governments of industrially developed countries, which are fully aware of their countries' comparative advantage in the field of science and technology, will tend to resist or try to thwart any international action that might endanger the competitiveness of their internationally active enterprises in crucial world markets, i.e., those with a high or above-average growth potential.[63] Thus, increased transfer of technology is not likely to be a method for "an improved distribution of technologies needed for increasing global welfare"; rather, technology transfer is more often than not subordinated to the worldwide sourcing and production strategies of transnational enterprises that are primarily based in the Western industrially developed countries.[64]

Without denying these built-in qualitative aspects of technology transfer by transnational enterprises, another position detects more decision and action latitude on the part of developing countries than the adherents of the first position would concede. Here it is argued that the competition among industrially developed countries and their transnational enterprises for markets and investment opportunities in the Third World provides the governments of developing countries with a bargaining power the full and determined use of which would allow them to achieve simultaneously an increased inflow of technologies as well as the establishment of both qualitative and quantitative controls over technology imports.[65] This argument could be thought of as applying best, at least for the foreseeable future, to the major oil-exporting and the newly industrializing countries. They seem to satisfy some of the conditions that lay the ground for a "self-reliance of judgment"[66] in the selection, acquisition, and adaptation of externally supplied technologies; however, even self-reliance is no insurance against making the wrong decisions.

Decision making with respect to the development and transfer of technology can be left to market processes or can be incorporated, at least in part, into public policymaking. In the context of mobilizing and applying science and technology for development some government representatives of major industrially developed countries still "believe that technology transfer . . . will contribute to meeting human needs and developing human capacities and to upward mobility through the growth of indigenous technical and managerial skills."[67] But even spokesmen for transnational enterprises concede that transfer of technology on their part bears little or no relationship to the promotion of indigenous technological capabilities in developing countries.[68] Thus, there has been growing support for the view that the development and transfer of technology is a legitimate concern of public policymaking of prime importance in developing countries (and, of course, in industrially developed countries, too). However, the question then arises what criteria should guide public policymaking in this field.

The discussion about this question has proceeded, by and large, under the label of "appropriate technology."[69] The basic issue has been whether and to what extent technology transfer might be harmful to the development of developing countries inasmuch as "the techniques from the advanced countries have inappropriate characteristics for underdeveloped countries."[70] The basic argument about the potentially inappropriate nature of advanced countries' technologies for developing countries derives from the differences in their respective physical and institutional environments as well as factor endowments. Put differently, advanced countries' technologies have been developed to suit the production systems, income levels, infrastructures, and so on of these countries. Their transfer to developing countries without adaptation and transformation cannot but lead to structural distortions in the receiving countries' economies and societies. Moreover, if the advanced countries' conditions cannot be reproduced in the receiving sector of the developing country, i.e., if, as is not at all unlikely, the structural distortions do not fully materialize, the technology transfer may even turn out to be inefficient because the output achieved in the developing country would fall below developed countries' standards.

What, then, would be appropriate technologies for developing countries? And how would technology transfer be affected by public policymaking guided by appropriate technology criteria? There have been many attempts at defining appropriate technology, as is reflected in the wide variety of terms covering the same idea: "intermediate technology," "village technology," "labour-intensive technology," "progressive technology," "Third World technology," "alternative technology," and so forth.[71] However, "all these have in common the idea that the Third World needs an alternative technology to that of the advanced countries."[72] Before turning to the distinguishing characteristics of appropriate technology one is reminded that "two big 'families' in appropriate technology"[73] have to be kept apart, one that relates to the modern sector (of industry, services, etc.), the other to the traditional and informal sectors.

Frances Stewart has elaborated a detailed scheme of requirements for more appropriate technologies for each of the two sectors in developing countries.[74] The six principal requirements relate to investment, scale of production units and their location, skills, material inputs, and products. As to investment requirements, the choice of more appropriate technologies consists in reducing the investment/labour ratio in the modern sector while increasing it in the traditional and informal sector. The size requirements of more appropriate technologies invariably necessitate a small or at least smaller scale of production units than would be warranted by advanced countries' technologies. Appropriate technologies would also favour locating production units in rural areas without, however, neglecting urban areas altogether, particularly as modern-sector technologies are already in use there. More appropriate technologies would minimize skill requirements for making, operating, maintaining, and repairing machines as well as for organizing input and marketing processes; obviously, skill requirements in the existing modern sector would be absolutely higher than in the traditional and informal sector, but they need not be excessively higher. Locally available and generated material inputs (natural resources and machinery) will also be preferred by applying more appropriate technologies to producing goods and services in developing countries. Finally, more appropriate technologies will lead to the production of goods and services designed for prevailing, i.e., low income levels in developing countries or of intermediate products that are capable of being processed further in Third World localities. This emphasis on more appropriate technologies for developing countries in both their existing modern and their traditional and informal sectors aims at breaking "the circle of continued dependence on foreign technology inhibiting the development of an autonomous capacity for technological change."[75]

Regardless of agreement or disagreement with the position that a more autonomous and balanced development of developing countries requires technologies other than those of the advanced countries, the question arises whether such technologies are indeed available and where they would come from. One answer is that there is an as yet untapped potential of traditional technologies, in both advanced and developing countries, that could be recovered and upgraded.[76] This approach could be particularly fruitful for the traditional and informal sector of developing countries' economies. However, appropriate technology must not be equated with traditional, outdated, or second-hand technology. The generation of efficient appropriate technologies certainly requires a major specific R&D effort, as "appropriate techniques have inevitably suffered, as to their existence, their number and their productivity, from their historical neglect in the development of technology."[77] This R&D effort need not be confined to developing countries particularly insofar as the modern sector is concerned. The need for more appropriate technologies in this sector may not be altogether dissimilar in both industrially developed and developing countries. Thus, advanced countries' R&D activities for more appropriate technologies in the modern sector need not, by their very nature, be irrelevant to developing countries; and the transfer of such technologies could not be said to entail, by necessity, structural distortions. It follows, then, that "selective delinking"[78] of developing countries from advanced countries might be called for to break away from a pattern of inappropriate technology transfer; yet, "selective coupling" might be equally valid as a strategic orientation toward the development and transfer of appropriate technologies. The difficulties inherent in following through with an approach termed here "selective coupling" cannot be ignored. But public policymaking in both industrially developed and developing countries supporting "selective coupling" for the mutually beneficial development and sharing of appropriate technologies must be seen as a challenge.

C. *International Scientific-Technological Cooperation and the Problem of Asymmetries*

International scientific-technological cooperation, be it bilaterally or multilaterally organized, has spread considerably over the last decades.[79] Yet, the available evidence suggests that the patterns of cooperation largely reflect, and may even tend to reproduce, the gap in scientific-technological capabilities between industrially developed and developing countries that was described in section 6A. Put differently, cooperation among developed countries is not only more frequent than cooperation between industrially developed and developing countries, let alone among developing countries themselves; it also appears that the sharing of results of scientific-technological cooperation follows the same stratification pattern. In the United Nations concepts of development, considerable emphasis has been placed upon increasing scientific-technological cooperation both between industrially developed and developing countries and among developing countries. This latter dimension of international cooperation for development, neglected for too long, has been, in the meantime, the subject of another global ad hoc conference of the United Nations system organized by the United Nations Development Programme (UNDP), the United Nations Conference on Technical Co-operation among Developing Countries (TCDC), and has resulted in promising follow-up activities.[80] Again, the question needs further probing: Can we expect that these objectives will be achieved and, if achieved, that they will significantly reduce the scientific-technological dependence of Third World countries?

If we accept a notion of cooperation that implies comparability of effort and commonality of interests as well as a proportionate sharing of costs and benefits, then scientific-technological cooperation among developing countries might hold the greatest prospects for advancing their autonomous development in this field. Aside from this theoretical consideration, practical evidence and projections of past trends justify the expectations that TCDC in the field of science and technology will become a powerful factor in the development process of the Third World and that, eventually, it will also be conducive to lessening Third World scientific-technological dependence on the advanced countries.[81]

This is not to argue that TCDC in the field of science and technology will evolve without problems. But it may be seriously misjudging the potential of TCDC to characterize it as "a position of withdrawal rather than of progress."[82] The crucial problems facing scientific-technological cooperation among developing countries stem from their colonial or semicolonial past and from the structural heterogeneity within and among them. One obvious problem derives from the wide discrepancies in scientific-technological capabilities among developing countries. Thus, structural asymmetries may stand in the way of authentic cooperation between Third World countries at different stages of scientific-technological development. Another difficult problem arises from the historical and still widely shared orientations of the scientific-technological communities in many developing countries toward former colonial or semicolonial powers. Thus, traditional North-South linkages and the old and new incentives supporting them compete with efforts at strengthening South-South linkages. A third problem has to do with political regime instabilities and incompatibilities in the Third World. Scientific-technological cooperation requires long-term commitments that must not be endangered by frequent political upheavals that, occasionally, put the very fabric of scientific-technological institutions and processes in jeopardy. However, although these problems are not insoluble, they lend weight to the prediction that increases in scientific-technological cooperation among developing countries are likely to concentrate on the more advanced and more stable and those countries with a stronger cultural identity of their own.

Turning to the goal of increasing the scientific-technological cooperation between industrially developed and developing countries, one should not lose sight of the danger that "cooperation" may turn out to represent "old wine in new bottles," i.e., a relationship of domination and paternalism, if the preconditions for cooperation are not present. For cooperation and the persistence of basic structural asymmetries must be considered as mutually exclusive unless special compensations in favour of the less advanced partner are built into the cooperation arrangement. Therefore, if authentic scientific-technological cooperation between industrially developed and developing countries is to be promoted, one has to look for programmes and projects that are capable of satisfying the preconditions of cooperation as set forth, for instance, in the "Pugwash Guidelines for International Scientific Co-operation for Development."[83]

As an illustration, two possibilities come to mind. First, one could focus on areas for scientific-technological cooperation in which developing countries have already reached a sufficient level of competence and expertise. Second, in other areas cooperative programmes and projects would be linked to promotional schemes that provide for raising the level of competence of participating scientists and technologists from developing countries up to that of their partners from industrially developed countries during the course of implementing the programme or project. In any event, there is no lack of ideas for cooperation, as indicated by the number and scope of international cooperation agreements in recent years.[84] However, if scientific-technological cooperation, particularly that between industrially developed and developing countries, is to contribute to development in the Third World of the kind alluded to in section 2, it will do little good to achieve more cooperation per se without paying enough attention to its social and political implications.

8. Dimensions of Conflict in the Issue Area of Science and Technology for Development

Cutting across the formulation of principles and the setting of goals that are indicative of the envisioned content of a new international order for science and technology based on a relatively broad consensus, a number of basic policy issues emerge that represent the contentious side of this new international order. These include

- the scope and intensity of state intervention in the process of promoting and applying science and technology to development;

- the methods of sharing in the world's scientific and technological potential; and

- the provision and allocation of financial resources for the development of science and technology in the developing countries in particular.

A basic point of contention between developing countries (and the Socialist countries) and most industrially developed OECD countries in conceiving of new international orders relates to the *scope and intensity of state intervention* in social processes in general[85] and in the process of promoting and applying science and technology to development in particular. Generally speaking, most developing countries—at least at the level of programme formulation, if not in actual practice—tend to favour a definition of state functions that puts public authorities in direct charge of steering the development process. One fundamental rationale for this position is, of course, the notion of greater national autonomy through increased state power. In the sphere of science and technology this implies broad responsibilities of the state for all aspects of science and technology policymaking, resource mobilization, and infrastructure building, and for closely monitoring and even controlling all scientific-technological transactions with the outside. Accordingly, nonstate entities (business, academic, professional, etc.) are expected to be subordinated and to conform to state guidance. Many industrially developed OECD countries object to this broad concept of state responsibility, particularly in the field of science and technology. Although certainly not denying the reality and necessity of state intervention in support of and to regulate the development of science and technology and its applications, they tend to insist on safeguarding the relative autonomy of the private sector both at home and abroad as a means of optimizing the growth and practical use of the available scientific-technological potential.

Considerable disagreement between developing and industrially developed countries also pervades the debate over the *methods of sharing in the world's scientific and technological potential.*[86] In a nutshell formulation: The developing countries demand the greatest freedom possible in their access to existing scientific and technological information, in utilizing existing productive technologies, be they proprietary or in the public domain, and in disposing of their scientific-technological manpower while retaining full autonomy in regulating the inflow of information and technology and the outflow of skilled personnel. Most industrially developed countries, however, although they do not object to the aspirations of developing countries in the areas of access to, and utilization of, existing scientific-technological potential at home and abroad, tend to point to the legitimate protection of authors' and inventors' rights and insist on the obligation of receiving countries to negotiate the terms of transfer agreements rather than imposing them by political-administrative fiat. A comparable attitude prevails on the issue of the "brain drain." Although they agree that it is desirable that highly skilled personnel stay in, or return to, their countries of origin, industrially developed OECD countries emphasize the human rights aspect of this issue and perceive it as a structural problem that can be solved only by improving the status and working conditions and by liberalizing the social and political environment of scientists and technologists.

The provision and allocation of financial resources for the development of science and technology in developing countries[87] constitutes another divisive issue in the struggle over a new international order for science and technology. It is generally agreed that the financing of science and technology development in Third World countries should come primarily from their own economic surplus. However, for most developing countries, although to varying degrees, i.e., according to their individual levels of development, this surplus is not sufficient to provide adequate financing for science and technology development. Thus, it is widely recognized that external funds, i.e., a transfer of financial resources from industrially developed countries, are indeed required to promote science and technology development in Third World countries to the level of self-sustaining growth. Relatively sharp differences between developing and industrially developed countries have arisen, however, over the criteria for, and the modalities and mechanisms of, such financing. The most far-reaching proposal is to tie the transfer of financial resources to indicators of technological dependence so that the flow of resources would not only be automatic and predictable, at least in the medium term, but would also vary according to changes in the patterns of technological dependence over time. There would thus be built-in incentive for industrially developed countries to commit themselves to reducing the technological dependence of developing countries or to accept the long-term prospects of compensation payments. Obviously, such an arrangement does not appeal to the industrially developed countries for a variety of reasons, even though some have already begun to reconsider their basic position toward the general issue of development finance.

In any case, most industrially developed countries do not resist, or would even welcome, an expansion of financial assistance for science and technology development in Third World countries on a bilateral basis or through established multilateral channels, such as the World Bank and its associated institutions, regional development banks, and UNDP. Furthermore, they seem to be ready for a series of more specific improvements in development finance, such as untying of aid, local cost financing, use of local experts and consultants, and so on. Thus, although compromise solutions on this and the other conflict dimensions do not seem impossible, there is no indication that the road toward a new international order for science and technology will not remain tortuous indeed.

9. Conclusion

A new international order for science and technology will be a system of limited collective state interventionism based on an enhanced state responsibility for the steering of scientific and technological change, especially in developing countries. It is part of an emerging global compact helped along by a variety of concerted multilateral action mechanisms (including UNCSTD and other development-oriented efforts of the United Nations system) that seek to keep a delicate balance between the demands of the developing countries for both respecting and strengthening their autonomy, and the dominant interests of the industrially developed countries' leadership groups in safeguarding and furthering their access to the resources and markets of developing countries, which is being perceived as fundamental to their stability and welfare. The politicization of development activities and international economic relations, which the United Nations system has no doubt facilitated, serves to compensate the developing countries for the structural weakness in their dealings with the industrially developed countries. By the same token, however, it also exerts a degree of consensus pressure on the developing countries that prevents them from exploiting their numerical majority and keeps the developed countries committed to a policy of working through the United Nations system as suiting their own vital long-term interests.

Notes

1. General Assembly Resolutions 3201 and 3202 (S-VI) in *General Assembly Official Records (GAOR), 6th Special Session, Supplement No. 1* (A/9559), pp. 3–12. Equally important in this context are General Assembly Resolutions 3281 (XXIX) in *GAOR, 29th Session, Supplement No. 31* (A/9631), pp. 50–55 ("Charter of Economic Rights and Duties of States") and 3362 (S-VII) in *GAOR, 7th Special Session, Supplement No. 1* (A/10301), pp. 3–9 ("Development and International Economic Co-operation").

2. For the state of negotiations on a new international maritime order see the *Informal Composite Negotiating Text/Revision 1 of 28 April 1979* (Doc. A/Conf. 62/WP.10/Rev. 1).

3. On this dimension of the debate over new international orders see the report of the UNESCO-sponsored MacBride Commission, International Commission for the Study of Communication Problems, *Final Report* (Provisional Version) (Paris: UNESCO, 1979).

4. According to a standard definition,

> the concept of "science and technology" applies to the following basic activities:
>
> (I) *scientific and technological research* (R) or the study, experimentation, conceptualization and theory-testing involved in making discoveries or developing new applications;
> (II) *experimental development* (D), which consists in the processes of adaptation, testing and refinement which lead to practical applicability;
> (III) *scientific and technological services* (STS) representing a mixed group of activities crucial to the progress of research and to the practical application of science and technology. These services collect, process and disseminate the scientific and technological information needed for such purposes;
> (IV) *innovation* or the development of a new product or process with a view to ensuring that fresh ideas and inventions are used effectively in the national economy.

An Introduction to Policy Analysis in Science and Technology, Science Policy Studies and Documents, No. 46 (Paris: UNESCO, 1979), pp. 7–8.

5. The phenomenon I am referring to is aptly described in a recent newspaper article: "The danger . . . is that in the quest for solutions to what are fundamentally political and social problems, the gusher of interest in science might simply lead to swapping one set of witch doctors for another." Daniel S. Greenberg, "Science: Society's New Crutch," *International Herald Tribune*, 12 January 1979, p. 4.

6. The most distinctive characteristic of the technocratic approach to social problem solving consists of eliminating the difference between praxis and technique: The claim to knowing about the requirements and constraints of a hypostasized logic of progress tends to preempt the possibilities of freely discussing and choosing among socially valued goals. For an in-depth theoretical analysis see Jürgen Habermas, *Technik und Wissenschaft als "Ideologie"* (Frankfurt: Suhrkamp, 1968), especially pp. 48 *et seq.*

7. The case of hunger and malnutrition is used for illustrative purposes only. It has been chosen because it highlights most vividly the fallacy of believing in technological fixes for what are, more often than not, basically sociopolitical problems. For a general statement see Susan George (Rapporteur), "An Issues Paper—Contributed by the Food Study Group of the

GPID Project," Tokyo, United Nations University, 1979 (HSDRGPID-2/UNUP-54).

8. Jacques Chonchol, "L'alimentation mondiale: L'echec des solutions productivistes," *IFDA Dossier*, No. 13 (November 1979), p. 43. The 1975 Dag Hammarskjöld Report analyzed the situation in the following way:

> Hunger and malnutrition are indeed due to the fact that the poor are deprived of the means either to produce or to purchase their food, the socio-economic mechanisms being so organized as to ensure that the lion's share goes to the rich and the powerful. The satisfaction of the need for food and its production cannot, therefore, be set apart from a transformation of political and socio-economic structures. . . . the dependence of many Third World countries has been increased by technological solutions isolated from the economic, social and ecological context to which they apply; the "green revolution," which entails the use of large quantities of chemical fertilizers, is a case in point.

See "What Now? Another Development," *Development Dialogue*, No. 1/2, 1975, pp. 30–31.

9. This argument has found recognition by the secretary-general of the special FAO Conference on Agrarian Reform and Rural Development, Hernan Santa Cruz, who wrote that the causes of rural poverty include, *inter alia*, "the inertia of obsolete socio-economic structures and the resistance to change by privileged groups. . . ." See his article "Land for the Lost," *Development Forum*, Vol. 6, no. 4 (May 1979), p. 1.

10. Cf. World Conference on Agrarian Reform and Rural Development, Rome, 12–20 July 1979, *Report* (Rome: FAO, 1979). Back in 1969 the Pearson Commission had already complained that "in many countries, land reform and rural development have not yet received the attention they require." See *Partners in Development. Report of the Commission on International Development* (Lester B. Pearson, Chairman) (New York, Praeger Publishers, 1969), pp. 13, 62.

11. The preceding observations have been informed by several studies that seem to converge in their critical analyses of many past and contemporary policies to fight hunger and malnutrition: Rudolf Buntzel, "Umriss einer landwirtschaftlichen Entwicklungsstrategie für Afrika," in Alfred Schmidt (ed.), *Strategien gegen Unterentwicklung* (Frankfurt: Campus, 1976), pp. 199–214; Kurt Egger and Bernhard Glaeser, "Ideologiekritik der Grünen Revolution," in *Technologie und Politik*, Vol. 1 (Reinbek b. Hamburg: Rowohlt, 1975), pp. 135–155; and Dieter Senghaas, *Weltwirtschaftsordnung und Entwicklungspolitik* (Frankfurt: Suhrkamp, 1977), pp. 189–202.

12. General Assembly Resolution 2626 (XXV), in *GAOR, 25th Session, Supplement No. 28* (A/8028), pp. 39–49 ("International Development Strategy for the Second United Nations Development Decade").

13. *World Plan of Action for the Application of Science and Technology to Development*. Prepared by the Advisory Committee on the Application of

Science and Technology to Development for the Second United Nations Development Decade (New York: United Nations, 1971) (E/4962/Rev. 2; Sales No.: E. 71. II.A. 18).

14. See note 1 above.

15. John M. Logsdon and Mary M. Allen, *Science and Technology in United Nations Conferences. A Report for the UN Office for Science and Technology* (Washington, D.C.: Graduate Programme in Science, Technology and Public Policy, George Washington University, January 1978), especially pp. 7-8. A factual account of the prominence attributed to science and technology in recent global conferences of the United Nations system can be found in a document requested by the Preparatory Committee for the United Nations Conference on Science and Technology for Development (UNCSTD): *Review Paper on Progress in the Implementation of Recommendations of Recent United Nations Conferences Highlighting the Role of Science and Technology for Development* (A/Conf. 81/PC/33 and Add. 1 and Add. 2).

16. *Report of the United Nations Conference on Science and Technology for Development*, Vienna, 20-31 August 1979 (New York, United Nations, 1979) (A/Conf. 81/16 and Corr. 1; Sales No.: E. 79.I.21).

17. The concepts of technological optimism and pessimism are adapted from Thomas G. Weiss and Robert S. Jordan, *The World Food Conference and Global Problem Solving* (New York: Praeger Publishers, 1976), pp. 139-140.

18. Johan Galtung, "Towards a New International Technological Order," *Alternatives*, Vol. 4, no. 3 (January 1979), pp. 278-279.

19. Ibid., p. 280.

20. Cf. "Science and Technology—Promises and Threats," in Paul Abrecht (ed.), *Faith, Science and the Future* (Geneva: World Council of Churches, 1978), pp. 21-23.

21. Denis Goulet, *The Uncertain Promise. Value Conflicts in Technology Transfer* (New York: IDOC/North America, 1977), p. 17.

22. Cf. "Science and Technology in the Struggle for a Just, Participatory and Sustainable Society," op. cit. (note 20), pp. 1-7.

23. "Toward a New International Development Strategy: The Scheveningen Report," *Development Dialogue* No. 1, 1980, pp. 55-67.

24. See, for example, *Economic and Social Consequences of the Arms Race and of Military Expenditures*, Updated Report of the Secretary-General (New York: United Nations, 1978) (A/32/88/Rev. 1; Sales No.: E.78.IX.1), pp. 28 *et seq.*, 46 *et seq.*

25. This approach as applied to the technology sector is exemplified in the writings of Francisco Sagasti (*Technology, Planning, and Self-Reliant Development. A Latin American View* [New York, Praeger Publishers, 1979] especially Chapter 1), and of Frances Stewart (*Technology and Underdevelopment*, 2nd ed. [London: Macmillan, 1978], especially Chapter 5).

26. Cf. the paper sponsored by the UNCTAD secretariat, "Technological Transformation of the Third World—Issues for Action in the 1980s," Geneva, 1978 (UNCTAD/TT/9).

27. A typical example of this approach can be found in a working paper by the United Nations Office for Science and Technology, "Interactions Between Science and Technology and Long-Term Global Problems: An International Policy Perspective," prepared for the International Colloquium on Science, Technology and Development: Needs, Challenges and Limitations, Vienna, 13–17 August 1979 (ACAST/COLL/WG.XIV/WP. 16). Cf. also "Mobilizing Technology for World Development: Report of the Jamaica Symposium," in Jairam Ramesh and Charles Weiss, Jr. (eds.), *Mobilizing Technology for World Development* (New York, Praeger Publishers, 1979), pp. 48 *et seq.*

28. *Report of the Working Group on Interrelations Between Science and Technology and Longer-Term Global Problems*, International Colloquium on Science, Technology and Development: Needs, Challenges and Limitations, Vienna, 13–17 August 1979 (ACAST/COLL/WG.XIV/REP.14), p. 1.

29. "Interactions Between Science and Technology and Long-Term Global Problems," p. 4.

30. Ibid., p. 2.

31. Ibid., p. 3.

32. A succinct summary of this approach was given by Ann Mattis ("Science and Technology for Self-Reliant Development," *IFDA Dossier* No. 4 (February 1979), p. 5), who wrote:

> the problem could be said to lie in the fact that S&T have escaped from social control in industrialized and Third World countries alike. The potential users of a key instrument of the development process are not in control of that instrument. For this reason, the Conference [UNCSTD] must address itself to the fundamental issues of social control over science and technology and use of S&T for meaningful development. These however are not questions which should be the sole domain of scientists, technologists and diplomats. These must necessarily be citizen issues.

33. "What Now?" op. cit. (note 8), p. 93 (emphasis added).

34. For an analysis of the relative success that each of these competing conceptualizations had during UNCSTD, see Chapter 6.

35. Jan Annerstedt, *A Survey of World Research and Development Efforts* (Paris: OECD Development Centre and Roskilde: Institute of Economics and Planning, Roskilde University Centre, July 1979), p. 5. Cf. also, by the same author, "Das wissenschaftliche und technische Potential der Welt," *Neue Entwicklungspolitik* (Vienna) Vol. 5, no. 2 (1979), pp. 13–23.

36. On these indicators data could be collected across the largest number of nations. Annerstedt's findings are essentially corroborated by UNCTAD studies. However, where they deviate from one another in various details the discrepancies seem to be due to the smaller data base of the UNCTAD

studies. Cf. the background document for UNCTAD V, *Towards the Technological Transformation of the Developing Countries* (TD/238), especially pp. 25–30. On another indicator, the number of patents issued, Frances Stewart reported that "only 6 per cent of the estimated 3 1/2 million patents in existence in 1972 were granted by developing countries, and less than one-sixth of that total were owned by nationals of developing countries." *Technology and Underdevelopment*, op. cit. (note 25), p. 119.

37. Annerstedt, *a Survey*, op. cit. (note 35), p. 6.

38. Ibid., p. 21. In referring to these data I am, of course, aware of the obvious qualification that more R&D personnel does not automatically translate into higher R&D output and its absorption by the productive sector.

39. The following paragraphs paraphrase and summarize information spread throughout Annerstedt, *Survey* and "Das wissenschaftliche und technische Potential der Welt."

40. Stewart, op. cit. (note 25), pp. 123–138.

41. Cf. in particular the UNCTAD secretariat's study, *Major Issues Arising from the Transfer of Technology to Developing Countries* (New York, United Nations, 1975) (TD/B/AC.11/10/Rev. 2; Sales No.: E. 75.II.D.2), pp. 25–30. Direct costs of overt technology transfer include payments for patents, licenses, know-how, trademarks, management, and technical fees. See also Surendra Patel, "The Technological Dependence of Developing Countries," *UNITAR News*, Vol. 6, no. 4 (1974), pp. 23–25.

42. Definition taken from *Major Issues*, op. cit. (note 41), p. 27.

43. Stewart, op.cit. (note 25), p. 124.

44. Ibid., p. 125.

45. Ibid., p. 131.

46. On the state of negotiations on a code of conduct for the transfer of technology see the UNCTAD documents, *Draft International Code of Conduct on the Transfer of Technology* (as of 16 November 1979) (TD/CODE TOT/20), and *Selected Documents of the Second Session of the Conference* (TD/CODE TOT/21).

47. Stewart, op. cit. (note 25), pp. 131–132 and Chapters 3 and 4 *passim*; cf. also section 7B below.

48. The literature on appropriate technology has become too vast to be referenced. A fairly representative publication would be A. S. Bhalla (ed.), *Towards Global Action for Appropriate Technology* (Oxford: Pergamon, 1979).

49. Stewart, op. cit. (note 25), p. 132.

50. Ibid., p. 133. An interesting case study in this regard is G. S. Aurora and Ward Morehouse, "The Dilemma of Technological Choice in India: The Case of the Small Tractor," *Minerva*, Vol. 12, no. 4 (October 1974), pp. 433–458.

51. This point has been elaborated in some detail by Hollis Chenery et al., *Redistribution with Growth* (Oxford: Oxford University Press, 1974).

52. Cf. the General Assembly Resolution on the "International Development Strategy for the Second United Nations Development Decade" (note 12), paragraph 63.

53. On the multilateralization of development aid see Joachim Betz, *Die Internationalisierung der Entwicklungschilfe* (Baden-Baden: Nomos, 1978).

54. Cf., for instance, Francisco Sagasti, op. cit. (note 25), pp. 10–11.

55. See note 12.

56. Cf. *Report of the Second General Conference of the United Nations Industrial Development Organization*, Lima, Peru, 12–26 March 1975 (ID/CONF.3/31). For the text of the Lima Declaration and Plan of Action on Industrial Development and Co-operation, see pp. 44–63.

57. See "What Now?" op. cit. (note 8) p. 108.

58. The stepped-up activities of the Socialist International in recent years, attempting to involve reformist regimes in the Third World in its work, are a case in point.

59. Cf. the United Nations documents mentioned in notes 1 and 12.

60. See Dieter Ernst, "International Transfer of Technology, Technological Dependence and Underdevelopment: Key Issues," in Dieter Ernst (ed.), *The New International Division of Labour, Technology, and Underdevelopment* (Frankfurt: Campus, 1980), especially pp. 17–19, as well as Carlos Contreras, *Transferencia de tecnologia a paises en desarrollo* (Caracas: Instituto Latinoamericano de Investigaciones Sociales, n.d.), especially Chapters 1 and 2.

61. Galtung, op. cit. (note 18), p. 288.

62. Ernst, op. cit. (note 60), p. 37.

63. The protracted negotiations about a code of conduct on the transfer of technology (see note 46) reflect this fact quite clearly.

64. Ernst, op. cit. (note 60), pp. 23–24.

65. Cf. Constantine V. Vaitsos, "Government Policies for Bargaining with Transnational Enterprises in the Acquisition of Technology," in Ramesh and Weiss, op. cit. (note 27), pp. 98 *et seq.*

66. This phrase is borrowed from Ward Morehouse, *Science, Technology and the Global Equity Crisis. New Directions for United States Policy.* Occasional Paper 16. (Muscatine, Iowa: Stanley Foundation, 1978), p. 19.

67. See, for example, on behalf of the U.S. government, Joseph S. Nye, Jr., "Science and Technology – Technology Transfer Policies," *Department of State Bulletin*, Vol. 78, no. 2012 (1978), p. 40. The author was then Deputy to the Under-Secretary for Security Assistance, Science and Technology in the U.S. Department of State.

68. Cf. Business International, *Transfer of Technology. A Survey of Corporate Reaction to a Proposed Code* (Geneva, 1978), pp. 13–15, where the main reasons for transferring technologies by transnational corporations to developing countries are described. These are, not surprisingly, very mundane indeed.

69. Cf. again note 48 and section 6B(1) above.

70. Stewart, op cit. (note 25), p. 60.

71. Ibid., p. 96; and Donald D. Evans: "Appropriate Technology and Its Role in Development," in Donald D. Evans and Laurie Nogg Adler (eds.), *Appropriate Technology for Development: A Discussion and Case Histories* (Boulder, Colo., Westview Press, 1979), p. 45.

72. Stewart, op cit. (note 25), p. 96.

73. See N. Jéquier: "Appropriate Technology: Some Criteria," in Bhalla, op. cit. (note 48), p. 4.

74. For the following see Stewart, op. cit. (note 25), pp. 98–106.

75. Ward Morehouse and Jon Sigurdson, "Science and Technology and Poverty: Issues Underlying the 1979 UN Conference on Science and Technology for Development," *Bulletin of the Atomic Scientists*, Vol. 30, no. 10 (1978), p. 26.

76. An interesting example is the United Nations University's research programme described in *Sharing of Traditional Technology: Project Meeting Report*, Tokyo, September 1977 (Tokyo: United Nations University, 1978) (HSDPD-8/UNUP-9). On the recovery of traditional technologies cf. *Models for Rural Development* (London: Acton Society Trust in association with the Commonwealth Science Council, 1979).

77. Stewart, op. cit. (note 25), p. 109.

78. This is the key strategic concept in critiques of technological dependence; cf. Dieter Ernst, "Consequences for National Policies to Strengthen Technological Self-Reliance," in Ernst, op. cit. (note 60), pp. 608–611.

79. Cf. Jean Touscoz, *La coopération Scientifique internationale* (Paris: Editions Techniques et Economiques, 1973).

80. Cf. *Report of the United Nations Conference on Technical Cooperation Among Developing Countries*, Buenos Aires, 30 August – 12 September 1978 (New York, United Nations, 1978) (A/Conf.79/13/Rev. 1). On the follow-up activities see *TCDC News*, No. 6 (April – June 1980), which paid special attention to the 1980 review meetings.

81. For details see Chapter 3 by Ward Morehouse and David Chu.

82. See Ernst, op. cit. (note 60), p. 41.

83. Published in the *Pugwash Newsletter*, Vol. 16, no. 3 (January 1979), pp. 70–80.

84. Take, for instance, the case of the Federal Republic of Germany and the evidence in *Bundesbericht Forschung VI* (Bonn, 1979), pp. 32–34 and 359–379.

85. See on this point, for example, Ismail-Sabri Abdalla, "Secteur public et stratégie de dévéloppement," *IFDA Dossier*, No. 7, May 1979.

86. See, for example, I. Wesley-Tanaskovic, "Scientific and Technological Information Systems for Development," in Klaus-Heinrich Standke and M. Anandakrishnan (eds.), *Science, Technology and Society: Needs, Challenges and Limitations* (New York: Pergamon, 1980), pp. 418–456; H. P. Kunz-Hallstein, "Promoting Research and Development for

Solving Problems of Economic Underdevelopment: The Importance and Possibilities of Patent Protection," in K. Gottstein (ed.), *Science and Technology for Development* (Starnberg: Max Planck Institute and Tübingen: Institut für wissenschaftliche Zusammenarbeit, 1979), pp. 107–110; and William Glaser, *The Brain Drain: Emigration and Return* (New York: Pergamon Press for UNITAR, 1978).

87. In addition to Chapter 4 by Francisco Sagasti, see Horst P. Wiesebach, "Mobilization of Development Finance: Promises and Problems of Automaticity," *Development Dialogue*, No. 1, 1980, pp. 5–28; and Volker Rittberger and John Renninger (comp.), *Financial Arrangements for the Promotion of Science and Technology for Development.* Report of a UNITAR Seminar, Schloss Hernstein, Austria, 18–21 July 1979 (New York, 1980).

2
Technological Autonomy and Disassociation in the International System: An Alternative Economic and Political Strategy for National Development

Ward Morehouse

1. Introduction

As we have moved from one decade to another, the agenda of the international community has been crowded with opportunities to examine the maldistribution of the world's scientific and technological capacities and the implications of that maldistribution for realizing the overriding goals of the United Nations for a more peaceful and just world order. This agenda has included such events in 1979 and 1980 as UNCTAD V (UN Conference on Trade and Development), UNIDO III (UN Industrial Organization), the UN Conference on Science and Technology for Development (UNCSTD), and the General Assembly Special Session on the Third Development Decade. But alas, in spite of a torrent of words, little has changed, and the problem of global maldistribution of science and technology is still very much with us. The debate is far from over.

This chapter is intended to be a contribution to the continuing debate on alternative strategies for dealing meaningfully with the badly skewed distribution of the world's problem-solving capacities based on modern science and advanced technology. Most of the world is a scientific desert, with some 95 percent of global expenditure on research and development concentrated in a handful of industrialized countries. The critical issue, however, is not the extent of the maldistribution but what can be done about it at the national level and by the international community. Needless to say, there are no quick or easy solutions.

In analyzing the role of technology in economic and political change, it is important not to fall into the trap of technological determinism. Although technology is a significant variable in the process of social change, the key prior questions are economic and political. Technology is a lever, which in conjunction with other initiatives can lead to a more autonomous, less dependent role for communities, social groups, or whole nations dominated by others in the present economic and political system. That system is in turn sustained by the character and direction of technological change. This change is further influenced by and reinforces existing economic and political relationships and institutions in circular causation. It is this vicious circle that some such initiative as technological delinking is needed to break.

Changed patterns of international trade and production of capital goods are key elements in a strategy to strengthen technological autonomy, but the core of such an initiative is a local technological innovation effort geared to meeting local needs.

The gap in technology between industrially developed and developing countries is not simply the result of the former's spending more money on research and development (R&D) — although they do by a factor of 20 to 1 — but because of their capacity for utilizing their R&D through technological innovation. It is this capacity for utilization of new technology more responsive to local needs and resources that is the key to greater technological autonomy for developing countries.

The strategy of selective technological and economic delinking or disassociation from the global production system stands in sharp contrast with the prevailing "integrationist" school of world development. The latter school argues that the best way for developing countries to progress is to rely on the benefits of international cooperation. It is, of course, by now widely recognized that the benefits of such international cooperation flow very unevenly. This has led to calls for a New International Economic Order, in which the economic and political rules of the international system would be altered to make them less discriminatory to developing countries generally and the poorest countries particularly.

Proponents of selective disassociation argue that although it is vitally important to continue trying to change the rules of the game, the present configuration of economic and political power in the world makes it unlikely that significant changes in the rules will occur until that configuration is altered. For this configuration to be altered, developing countries will need greatly strengthened capacities for solving their own problems. And for some countries, at least, a measure of disengagement appears to be a more likely route toward that goal than continued dependence on external assistance, which has all too often in the past thirty years had the net effect of perpetuating, rather than diminishing, their dependence.

Selective delinking is a transitional strategy, leading to relinking with the international system in a more autonomous and less dependent relationship. There is solid historical evidence for the viability of such a strategy. Virtually all the modern industrial nations at one time or another went through a period of relative disengagement during which their ability to increase production and productivity spurted forward, permitting subsequent reengagement with the international system from a much stronger position.

But in the modern world, implementation of such a strategy remains elusive because most of the more recent cases of delinking were imposed externally rather than as the result of a country's own initiative. Delinking is strong economic and political medicine not often taken voluntarily by those whose long-term condition it would benefit.

But if the benefits of this strategy were better understood, perhaps resistance to it would be diminished. At least, consideration of such a strategy for national development among various alternatives should not be foreclosed before it has been examined. The purpose of this paper is not to assert that delinking is the only strategy of national development to the exclusion of all others but simply to urge that in weighing alternatives, delinking not be excluded.

2. The Basic Thesis

Disassociation is widely regarded as a quaint, implausible, unrealistic, or even dangerous thesis for the 1980s, even though it has solid historical precedents. In common with a number of other possibilities before the Third World in the 1980s, it is a "least worst" strategy not unlike Winston Churchill's definition of democracy as the worst form of government there is, except for all others.

Most policy initiatives in recent decades have assumed that the best way to attack inequities within and among nations through the international system has been to increase the flow of technology in the form of skills, knowledge, and hardware from North to South. Indeed, this was the animating idea behind the original initiative by the largest purveyor of aid to developing countries in the postwar era, beginning with U.S. President Truman's fourth point in his 1949 inaugural address. It has been the cornerstone of not only U.S. but other industrialized-country development policies, as well as those of multilateral institutions, ever since.

In view of the mounting evidence that such initiatives have had at best only limited impact on reducing (and may have even exacerbated in certain circumstances) inequities within and among nations, it is time to look at the reverse of this proposition, namely, the possible consequences of selective technological disassociation of North and South for the international system and for key nation-states in that system. What follows is only the most tentative and preliminary effort at such analysis; even more tentative and preliminary is the concluding discussion of how to implement such a strategy through concrete policy measures.

Perhaps the best way to open up this complex topic for further exploration is by setting forth four propositions, which might well form ,part of the agenda for the continuing debate on the role of science and technology in Third World development and North-South relations as we enter a new decade.[1]

A. Technology as an Instrument of Global Domination

As other forms of military and economic power of the major industrialized countries decline, technology is emerging as a significant and forceful instrument for maintaining existing dominance/dependence relations in the international economic and political system. ("Policy leverage" is the polite phrase used in policy documents of such countries.[2]) In political terms this is what international events such as UNCSTD, UNCTAD V, UNIDO III, and the UN General Assembly Special Session on the Third Development Decade are all about.

It is not necessary to subscribe to malevolent conspiracy theories to label the predominant structural characteristic of the international system as one of dominance and dependence. Industrialized countries and other instruments of technological domination like transnational corporations try to maintain their dominant positions not in order to deprive developing countries and exploit the poor, but rather to maximize their own interests, one of the results of which sometimes happens to be exploitation of others.

B. Technological Autonomy, Not Capacity, as the Goal of Developing Countries

If the first proposition is valid, then the goal of developing countries should be not to strengthen their technological capacity when that capacity is essentially derived from the industrialized countries. Instead it should be to increase their autonomous capacity for directing technological change toward meeting their own economic and social problems. Growing evidence suggests that increasing technological capacity along derivative lines all too often results in continuing dependence at qualitatively higher levels.[3] Technological autonomy is not, of course, autarky, but it does imply greater selectivity in and closer control of externally acquired technology.[4]

C. Selective Technological Disassociation as a Strategy for Achieving Greater Autonomy

If technological autonomy, not mere capacity, is the goal for developing countries and if, as a number of analyses suggest, it is very difficult to achieve autonomy in the face of increasing technology flows from industrially developed countries, then a strategy of *selective* disassociation may be for some countries a more certain and effective means of realizing significantly greater autonomy.[5] Selective delinking does not, of course, mean technological isolation of the Third World, which would obviously be foolish in view of the overwhelming dominance of the industrialized countries in virtually all areas of modern technology. It does mean carefully targeted technology acquisitions *on the initiative of developing countries*, rather than the present system of indiscriminate North-South technology flows, allegedly determined by the "market mechanism" (highly imperfect in international technology transactions), but in fact heavily influenced by developed-country governments and corporations through tax, trade, and development policies and marketing techniques designed to influence product choice. It is a transitional strategy, leading to subsequent relinking with the international system in a more autonomous relationship.

It also will require *far greater* efforts and *meaningful* scientific and technological cooperation among developing countries. Such a strategy might not have been possible in the past, but by the 1980s enough beachheads of technological competence, even in high-technology areas like nuclear energy, space, electronics, and computers, have been established in the Third World to make collective self-reliance a much more viable approach *if* developing countries are prepared to engage in genuine sharing among themselves.

D. Technological Autonomy Versus Dependence in Achieving Greater Equity Among and Within Nations

There are many variables involved in complex and dynamic inter-relationships in moving toward greater equity within nations, of which technology is only one. But if one measure of greater equity is the ability to meet the minimum material needs of a majority of the people on a self-sustaining basis, then most of those countries that have gone farthest down this road have one capacity in common—an increasingly autonomous capacity to create, acquire, adapt, and use technological solutions appropriate to their economic and social conditions.

This autonomous capacity is found in countries with very different natural resource endowments, population size, and political and economic systems. It is difficult to imagine two more sharply contrasting examples than the People's Republic of China and Taiwan. Yet each in its own way has markedly strengthened its technological autonomy while pursuing these goals. China has been following a predominantly closed, self-contained (but by no means technologically isolated) approach to development, certainly since the end of Soviet economic assistance in 1960. Taiwan, by contrast, has been following a more open strategy of export substitution, especially in the last fifteen years, although how many Taiwans with truly large economies the industrialized countries would tolerate is questionable.[6]

There is nonetheless no clear-cut set of causal relationships that should lead us to conclude that greater technological autonomy will necessarily lead to greater social equity within countries, although few developing countries have yet been able to meet the minimum needs of most of their people on a self-sustaining basis without a substantial measure of technological autonomy. Indeed, technology is typically a dependent variable in the complex calculus of social change, and economic, political, and cultural factors matter much more. On the other hand, greater technological autonomy appears to be an indispensable condition in realizing greater equity among nations. Countries that have limited problem-solving capacities of their own and that are largely dependent on others for providing the knowledge and skills needed to solve their problems are always going to be vulnerable, in the long if not the short run.

The vulnerability of developing countries, furthermore, is likely to grow in the 1980s as the trend toward automation of production technology and related technological changes in the industrialized countries accelerates.[7] Transnational corporations (TNCs), the principal carriers of industrial technology to the Third World, are concerned not only with maximizing profit but also with minimizing risk. Risk can be minimized by keeping as many elements in the production process as firmly under one's own control as possible. As developing countries become more assertive about their economic and political interests, the elements of both risk and cost for TNCs are likely to increase. This trend will only enhance the incentives to TNCs and their home countries to withdraw low-wage, labour-intensive production from developing countries. Instead, such production will be concentrated once again within the confines of the industrialized world, where it will be competitive with comparable production in developing countries because the automation of production technology will have substantially eliminated high-wage, industrialized-country workers.

As long as the capacity for technological innovation remains concentrated in the industrialized countries, developing countries will be vulnerable to technological choices made elsewhere. Industrialized-country choices are unlikely to be made in order to protect the interests of developing countries, especially where vital interests within industrialized countries would be adversely affected. On the contrary, the industrialized countries will make choices in technology, and more broadly in industrial and agricultural development policy, to maximize *their* autonomy and minimize *their* dependency on others, as long as the individual nation-state remains the key actor in the international political and economic system. Developing countries would be well advised to follow the same strategy.

3. The Problem of Terminology: Technological Autonomy and Disassociation Defined

A. *Technological Autonomy*

Technological autonomy or self-reliance is merely a shorthand way of referring to the autonomous capacity for creating, acquiring, adapting, and using technologies necessary to meet a country's economic and social needs. It is not an absolute but a relative condition and therefore involves an *increasing* capacity for autonomous decision making on the choice and use of technologies, including control of acquisition and use of foreign technologies and generation of indigenous technological solutions to a country's most urgent economic and social problems, especially those relating to meeting the minimum material needs of the people. Technology is here defined as the skills, knowledge, and tools needed to make useful things ("things" being understood to include services) and as the social process involved in making such things through the interaction and use of skills, knowledge, and tools. Technology thus becomes an important carrier of economic and political relationships within and among societies.

This autonomous capacity for technological development does *not* mean autarky or isolation, although it *may* involve selective disassociation from the international economic system for periods of time or sectors of economic activity while indigenous problem-solving capabilities are being tested and developed. Technological autonomy does not mean that a country must reinvent the wheel but rather that it should have the ability to do so if it had to. It clearly involves the capability to improve upon wheels invented elsewhere.

Technological autonomy is crucial to a country's capability for increasing its productive capacity. Without that capability, developing countries will never be able to meet the minimum material needs of their people, and if they cannot do that, they will always remain highly dependent actors in the world economic and political system, potential international basket cases that survive only at the sufferance of more affluent countries that are able to meet their own needs and help others unable to do so when it suits their purposes.[8]

Some find difficulty with the phrase "technological autonomy" because, taking the narrower definition of autonomy as self-rule, it appears to involve applying a political concept to a socioeconomic phenomenon. Decision making about the choice and use of technology can, of course, be and often is highly political when it influences access to and distribution of power within a society or among countries. But the broader meaning of autonomy as self-control is probably more relevant in the present instance, for it encompasses the more limited political connotation as well as the wider socioeconomic context where creation, acquisition, adaptation, and use of technology is a significant variable in the process of economic, social, and political change.

B. *Technological Disassociation*

Technological disassociation is also an elusive concept subject to misinterpretation. It is more a matter of style in a country's external technological relationships than a fixed position, although it is possible to conceive of a theoretical situation of complete delinking (and indeed some historical examples exist, such as isolated neolithic tribal groups in insular Southeast Asia or the interior of Africa in the nineteenth and first part of the twentieth centuries). It can be best defined by noting some of its characteristics, which are likely to be found in differing combinations and in greater or less measure in countries following a strategy of selective technological delinking. It should be seen as a transitional strategy, leading to relinking with the international system in a more autonomous and less dependent relationship.

It may involve a partial or complete prohibition on technology imports in selected sectors of economic activity (for example, nonessential consumer goods) for periods of time — a kind of "infant industries" approach to nurturing the local capacity for technological innovation and the ability to create new productive capacity. It is more likely to involve a mix of practices to unbundle technology acquisition. and installation—for example, use of indigenous design engineering and fabrication skills, together with some indigenously designed components, in creating a new production process within the country that also involves foreign components.[9] It may involve settling for less "efficient" production units (in terms of conventional Western economic analysis) based on indigenously designed and installed capacity where these generate substantially greater employment and are related to a country's efforts to satisfy its own minimum needs. It is likely to involve a search for more South-South technological and economic links, not to the exclusion of North-South ties but as a counterweight to lessen vulnerability in the latter instance, typically a more dependent and unequal relationship.

In general, there is likely to be more emphasis given to trying to find an internal solution to a given economic or social need before seeking an external one and to revitalizing the traditional technological base of the economy rather than pushing for its wholesale abandonment.[10] When technology is acquired externally, an effort will be made to maximize the arms-length character of the acquisition. As there are a variety of ways in which technology can be acquired, this can best be visualized as a continuum with maximum disembodiment and entangling conditions at one end and the reverse at the other. Perhaps the greatest arms-length transaction is one involving the purchase of blueprints and/or the direct hire of one's own nationals with industrialized-country experience in design engineering and industrial R&D. Intermediate points on the continuum involve such modes of acquisition as simple purchase of equipment, equipment purchase together with servicing and/or training of local personnel in its use and maintenance, and turnkey industrial projects, perhaps accompanied by a contract for foreign management for a period of time.

The most intimate mode of technology acquisition is usually considered to be direct foreign investment, in which the parent corporation retains control of all the know-how and marketing, not to mention the capacity to improve existing processes and create new ones, and simply integrates the developing-country subsidiary into a global production, marketing, and financing system that seeks to minimize costs and maximize profits for the parent company.[11] For the selective delinker, this last kind of acquisition is the least attractive, although such a country probably will follow a mix of different kinds of transactions, considering direct foreign investment if it is the only way to acquire a particular technology to which the country attaches a high priority.

The central thrust of a disassociation strategy is to disengage as much as possible from, and avoid becoming integrated any more than necessary into, the global production, marketing, and finance system dominated by the industrialized countries and TNCs head-quartered in these countries, for at least two reasons:

1. Industrialized countries and their TNCs will always act in the long run to maximize their interests, making Third World countries, as long as they are highly dependent actors in this system, very vulnerable. Increasing automation of production technology for risk minimization and profit maximization by TNCs in the 1980s already mentioned is a good case in point.
2. Integration of poor-country economies into this system has a sharply distorting effect on those countries, socially and technologically:
   a. Socially, through the now well-recognized phenomenon of the dual economy, in which 10–20 percent of the population participating in and served by the modern sector have the possibility of realizing some benefits from such integration, while the remaining 80–90 percent of the people stagnate in deprivation and misery.
   b. Technologically, through reliance on external sources of technology, leading to failure to develop and use local problem-solving capacities and to revitalize traditional technologies, which are simply overwhelmed by "superior" foreign technologies.

Whatever may be the specific mix a country elects to use in meeting its technological needs, disassociation is not going to involve total isolation. Even China during the height of the Cultural Revolution kept open a variety of channels to the rest of the world, including literature search services, which scanned the world's technological periodicals and translated those considered relevant to a particular sector of productive activity, and periodic industrial equipment exhibitions in China of foreign machinery (after which the Chinese would bargain hard to purchase as cheaply as possible the machinery so displayed). The Chinese example also underscores the transitional nature of a delinking strategy; since the middle 1970s, the Chinese have been actively relinking technologically with the rest of the world.

The emphasis is on *selectivity* and on the *qualitative* character of external technology acquisition where, all other things being equal, the selective delinker will seek to maximize its control of the technology thus acquired. The distribution of social costs and benefits within the country is a critical factor in making this selection. It is of necessity a long-term strategy, as it takes time to develop and deploy technological skills and acquire and diffuse knowledge within a country's production system, especially if that system has been functioning historically at a low level of productivity.

Technological disassociation is also not a recipe for heavy reliance on "village," "soft," or "intermediate" technologies, although improvements in such technologies will certainly be required to deal with the problems of rural poverty and of meeting the minimum needs of a majority of the people. It is rather a strategy of "walking on many legs." The problem for many developing countries at present, however, is that in their efforts to increase the productive capacity of their economies, they have been trying to walk on only one leg — namely, imported, capital-intensive technology for the modern sector of the economy, which serves only a small proportion of the total population.

A strategy of selective disassociation can be turned inside out and be seen as a strategy of "selective interdependence" (or more properly, at least at the beginning, "selective dependence"). This concept, which has been developed by Francisco Sagasti in relation to Latin America, contemplates that "the underdeveloped country will seek to concentrate its scientific and technological efforts in areas for which it already has relatively high competence, or can acquire it in the short-term, and in areas for which knowledge cannot (or should not) be imported."[12]

But whether the strategy is called "selective disassociation" or "selective interdependence," it involves hard choices in delaying growth for the sake of acquiring more autonomous technological capacities. Sagasti cited this example, based on his discussion with various ministers in a Latin American country:

> Should we build a turn-key fertilizer plant rather quickly (foreign exchange savings in fertilizer imports, support to agriculture, but no S and T learning) or should we unpackage the technology and do part

ourselves but more slowly and without guarantees (S and T learning but higher costs in foreign exchange and less support for agriculture and food). There is no easy way out.[13]

Finally, it should be recognized that technological disassociation is not the only route to achieving greater technological autonomy in the late twentieth century. The previously cited example of Taiwan, hardly disengaged economically or technologically over the past three decades, is a case in point. But it is one way, and for some countries unable to "kick the habit" of excessive technological dependence in any other manner, the only way—at least for a time.

4. Shadow and Substance in Technology Policies for Autonomy and Disassociation

Just as technology is only one variable among many affecting economic and social change, so also must technology policy be seen as an integral part of the domestic and foreign economic, social, and political policies of a given developing country. To argue that selective disengagement will lead to the technological transformation of a developing country and its emancipation from technological dependency on industrialized countries is fatuous in the absence of an accompanying social transformation. The major and most obvious examples of countries that have achieved, in varying degrees and with quite different social structures, a technological transformation in the twentieth century are Japan, the Soviet Union, and China. These countries have also experienced profound social transformation, two through political revolution and one through a more gradual process of social change in which cultural values proved to be highly persistent.[14]

It is this lack of effective integration of technology policy with overall economic, social, and political policies that has resulted in the ineffectiveness thus far of most policy instruments devised by developing countries to cope with technological dependence and enhance technological autonomy. These policy instruments, involving such devices as registries of foreign technology collaboration agreements, foreign investment control boards, and policies that assert priorities for foreign and indigenous technology, could be used as instruments for selective disengagement and enhancement of autonomy, but in fact do not, at least yet, appear to have had this kind of impact. India and Mexico are two examples; recent studies indicate that such policy measures have had little impact on continuation of an essentially dependent pattern of development, including a dominant role for foreign technology in key sectors of economic growth.[15]

This raises the haunting question of whether technology policy is a logical point for analysis or action, as it is clearly constrained by larger economic, social, and political policies and forces. It is no doubt partly an act of faith to assert that technology is a key variable in the process of societal response and adjustment to the rapidly changing social and physical environment in the world today. But if the trends of the last thirty years continue to the end of this century, technology – or more broadly, the capacity to generate, control, and use socially significant and instrumental knowledge – will be a critical, if not the critical, factor in how effectively countries cope with their own problems and with one another. In short, the old adage that control of knowledge is power bids fair to be an ever more accurate description of the future course of human affairs.[16]

5. The Adverse Consequences of Past Technology Flows and Their Implications for the Global Equity Crisis of the 1980s

The case for selective technological disassociation, which argues for a significantly different pattern in North-South relations during the balance of this century from that of the last thirty years, rests on two central propositions:

1. Substantial, although not universal, evidence indicates that all too often technology flows from North to South through public and private investment and technical assistance have done little to alleviate, and may in certain cases have exacerbated, inequities among and within nations.
2. Regardless of whether one accepts the foregoing proposition, the character of these flows and the patterns of economic growth that they engender and/or reinforce are totally inadequate for meeting the global equity crisis that will affect a sizable portion of the population of the world by the mid-1980s.

The 1970s found the world confronted with an equity crisis of truly compelling magnitude. In India – which probably contains a quarter or more of the one billion poorest people on earth – 40 percent of the population are believed to live below the Indian definition of the poverty line. This definition is calculated as the minimum necessary for a calorie intake sufficient to stay alive (currently about $50 a year) and is so spartan as to make anyone hovering near the poverty line in industrialized countries seem positively affluent.[17]

Income inequality is only one part of the equity crisis in poor countries. Another indicator of that crisis is unemployment, underemployment, and employment at very marginal rates of productivity. These are already at depressingly high levels; the number of landless and almost landless agricultural labourers in India for whom these conditions are particularly acute increased between 1961 and 1971 by 90 percent. But what lies ahead is far worse. From 1974 to 1984, the Indian male labour force alone will grow from 152 million to 196 million.[18] Comparable trends exist in other developing countries.

Generating employment on this scale, let alone achieving less skewed income distribution, is going to require radically different approaches to development from the "trickle down" strategies of the past quarter of a century, which are based on importation of capital-intensive technology from industrialized countries. The principal result of these strategies has been the creation of small urbanized pockets of prosperity (but still with plenty of abject poverty by advanced-country standards) in a sea of human misery and deprivation of even the most basic necessities of life. Few developing countries have been able to avoid this polarization between urban affluence and rural poverty, and there is even some evidence to suggest that the faster the economic growth, the more rapidly income inequality accelerates.

Since the mid-1960s, Brazil has been increasing its per capita GNP at the rate of 5 to 6 percent a year. From 1960 to 1970, while Brazil was aggressively importing industrialized-country technology in order to maximize export earnings, the share of the national income going to the top 5 percent of the population grew from 27 percent to 35 percent, according to the Brazilian government's own figures (private estimates give the top 5 percent as much as 46 percent of the national income).[19]

The point is subject to challenge, but I am convinced, as are many other students of the political economy of development, that the weight of the evidence supports the melancholy proposition that much (not all) foreign assistance, whether bilateral or multilateral, has certainly not improved and may have complicated the situation of a growing number of people in Asia, Africa, and Latin America.[20]

6. On Closing the Gap and Eradicating Poverty

The foregoing perspective is only partially at odds with one of the more authoritative recent global assessments of the record of and future prospects for economic growth in poor countries. Robert McNamara of the World Bank, who has of course an obvious interest in asserting that the Bank and comparable instruments of developed-country intervention in the Third World do produce results, stated in his September 1977 address to the Bank's Board of Governors that the average rate of growth of per capita income in developing countries grew at 3 percent a year from 1950 to 1975. He goes on to observe that this growth rate, impressive as it is, has not closed the gap between North and South (in fact it has widened) because " 'closing the gap' was never a realistic objective in the first place."[21]

Whether the growth rate in developing countries would have been greater if they had followed a more autonomous and less dependent pattern of economic and social development, emphasizing in the first instance meeting the minimum needs of a majority of their populations, and whether political and economic relations between a dominant North and a dependent South would have permitted a more autonomous pattern of change are not questions to which Mr. McNamara addressed himself. Nor can any one else ever know their answers, for they lie in the realm of speculation about what might have happened if things had been different. But the comparative picture of Indian and Chinese economic performance from 1960 until the mid-1970s, with Chinese industrial production suddenly shooting exponentially upward after more than a decade of relative technological autonomy and disengagement while Indian industry lumbers at a marginal rate of growth, titillates the imagination.[22]

Regardless of whether closing the gap between North and South was ever a realistic goal, Mr. McNamara would agree with what was said earlier in this chapter, that "developing countries over the past quarter century . . . have failed to eliminate, or even significantly reduce the massive poverty in their societies." He went on to observe:

> Unlike "closing the gap," reducing poverty is a realistic objective, indeed an absolutely essential one. And it is true that some developing countries have had ineffective policies in this matter. In retrospect, it is clear that too much confidence was based on the belief that rapid

economic growth would automatically result in the reduction of poverty—the so-called "trickle down" theory.[23]

What Mr. McNamara does not do is pose the question of whether the Bank, through its past policies, with their heavy emphasis on capital-intensive economic infrastructure projects designed and often built by industrialized-country experts and facilitating the kind of unbalanced economic growth that has failed to eliminate poverty, has in fact contributed to this situation. Other observers think the Bank has.[24]

But it is to McNamara's credit that he recognized the problem in the early 1970s and began a gradual reorientation of the Bank's work toward a direct attack on poverty in the Third World. How far this has progressed and whether the Bank has now passed the threshold from being part of the problem to being part of the solution is a subject worthy of debate. McNamara is quite convinced, however, that the World Bank now knows what to do about poverty, a somewhat alarming assertion in view of its track record, the substantial increase in the Bank's resources in general, and its effect in this direction that he projects over the next few years.

It would be wrong to sell the Bank short. It has access to substantial financial resources. McNamara's effort at reorientation has resulted in some unusual and relatively forward-looking camels getting their noses inside the World Bank tent. But the crucial question is whether the Bank itself is sufficiently autonomous to withstand real pressure from major donor countries, should their economic interests be adversely affected if the Bank began backing genuinely autonomous patterns of technological change in a major way. The recent case of the attempted (and partially successful) congressional prohibition on the use of U.S. funds for the World Bank to increase production of palm oil and other commodities haunts the future.[25]

It is no wonder that rhetorical calls for global wars on poverty by Northern leaders like McNamara lack credibility in the South. Carlos Diaz Alejandro of the Yale Economic Growth Center, who has made a nonradical case for selective economic delinking of North and South in the paper from which the following is quoted, puts it this way:

> Why not argue for a world-wide war on poverty and oppression? To those familiar with the rise and fall of the Alliance for Progress, the answer is obvious: No Northern government has both the credibility

and the resources to launch and lead such a program seriously and globally, not now and not in the 1980s. Proposed global bargains between Northern and Southern elites of the type: more aid in exchange for more redistribution and democracy are, at best, utopian, at worst, a new version of an old confidence game. The sincerity with which some Northern individuals deplore poverty and oppression in the South is to be respected, but it strains the imagination to believe that major Northern governments, particularly those of large countries, could place such concerns at the center of their policies toward the South in any sustainable fashion in the near future. Indeed, much remains to be done by some governments on their home front to eliminate poverty and injustice.[26]

The fate of the U.S. budget request for aid for fiscal year 1979—which proposed a substantial increase in funds devoted to meeting minimum human needs—is a melancholy affirmation of Diaz's central point. It was apparently shot down quite early in the budget cycle because, among other reasons, no one in his right mind would spend that much on disinterested humanitarianism.

So we are back again to where we started from. Developing countries must rely primarily on their own efforts to solve their problems. This argues for giving primary emphasis to increasing their own problem-solving capacity—i.e., enhancing their technological autonomy—which may in turn require a measure of selective disengagement in order that this autonomous capacity can be used, tested, and developed. Therefore, the modalities of external assistance, which will under the best of circumstances be marginal in size, become critical. If industrialized countries repress the emergence of this autonomous capacity by refusing to let local skills and knowledge be used or by insisting, as a condition of assistance, on continued access for direct foreign investment with maximum control of technological innovation, the game is hardly worth the candle.

Experience over the past thirty years does not offer much encouragement that these kinds of conditions will be relaxed unilaterally by the North, although shifts in bargaining positions and strengths will and indeed have brought about some significant changes, of which the Organization of Petroleum Exporting Countries (OPEC) seems to have been the most notable example. For the most part, however, North-South technology flows, whether in the public domain and supported through concessional economic aid or whether involving proprietary technology through licensing agreements, direct foreign investment, or other means, are likely to follow the same modalities as in the past, unless southern countries become sufficiently strong economically and politically to force changes. And this will be particularly difficult to do in the area of technology transfer because the North has such an overwhelmingly dominant position and the South such limited bargaining power.

7. The Unholy Alliance of Power, Privilege, and Technology

Underlying the irrelevance of most technological change in the Third World to the unmet social needs of the majority of the population is the unholy alliance between the rich countries, which collectively generate their own technology, and that 10–20 percent of the population of poor countries that have adopted rich-country consumption patterns. The needs of these largely urbanized upper-income groups in the Third World are thus well served by rich-country technology.[27]

Such groups feel little compulsion to break the links tying them to technological development in the advanced countries through investment, trade, and other economic and political connections that are the carriers of technology from North to South. These upper-income groups include most of those with political and economic power and the privileges that go with that power, and those in rich countries with whom they interact—political leaders and senior officials of banks, transnational corporations, and international agencies—also occupy positions of power and privilege in the industrialized countries. We are thus confronted with an alliance of those within rich and poor countries who determine the direction and character of technological change in order to maintain their own status and power and to serve in the first instance their own needs and only secondarily the needs of others.[28]

The systematic character of technological development, with productive activity based on one kind of technology frequently requiring complementary activities based on similar technologies, means that this unholy alliance becomes all the more pervasive in its impact on technological change in poor countries. Thus at the technical level, alternative technologies designed to meet the needs of those segments of the population now largely excluded from the benefits of technological change would also raise serious problems for the efficiency of the existing system of productive activity. The pervasive influence of this unholy alliance also spills over into the economic infrastructure and social services, which are required to operate a productive system based on advanced-country technology and thus absorb most of the resources available to build up and maintain such infrastructure and services.

In a similar manner, the influence of this unholy alliance pervades the political and economic systems of both developing and developed countries. As Frances Stewart put it in *Technology and Underdevelopment:*

> To the extent that governments consist of individuals who benefit from, and represent those who benefit from, the political economy in being, they may not wish or be able to challenge it. An alternative technology at a macro level involves an alternative political economy – a different distribution of the benefits of the economic system. Governments which have developed in one system may not be powerful enough to choose an alternative system. . . .
>
> The effective pursuit of an alternative appropriate technology would threaten interests in the advanced countries . . . who are currently benefiting from the use of advanced-country technology in developing countries. The continued use of advanced-country technology is at the heart of the continued dependence of the poor countries. It maintains the advanced countries' lead in technology and therefore permits, indeed necessitates, the continued sale of technology, goods and managerial services to poor countries, on terms favorable to the rich countries.[29]

In view of these circumstances, the choice of technologies made by developing countries is not surprising. Stewart argued that more choices exist than technological determinists appear to believe, but the choice is narrower than many economists have assumed. Because of the concentration of most of the world's research and development in the industrialized countries on industrialized-country problems, moreover, little effort is made to generate and improve alternative technologies that might be more suited to the social needs and factor endowments of Third World countries. The productivity of these alternative technologies is typically rather low and the range of products for which they can be used is limited.[30]

But if developing countries do have some choice, however limited (and of course they could have more choice if there were a significant shift in the allocation of the world's R&D resources to the problem of the poor majorities in poor countries), Stewart contended that determinants of that choice, which she calls selection mechanisms, become critical. Her analysis of technological choice, which is supported by my work and that of others on the political economy of science and technology for development, indicates a much more complicated process than is usually assumed by many development economists, with relative prices being just one of the criteria, often only of minor significance.[31] The markets being served, the type of product required, international trading patterns, and local income distribution, not to mention policies toward advertising and product standards and the availability of resources (including knowledge of alternatives), all play a role.

The pervasive and interlocking character of this unholy alliance limits the extent to which developing-country governments are able to follow alternative paths even if they should wish to. Many selection mechanisms are the outcome of the economic system and type of technology already being used. To quote Frances Stewart again:

> A system based on the use of advanced-country technology tends to generate high incomes among those employed with the technology, providing markets for the goods the technology produces. . . . Relative prices and availability of resources are also in part outcomes of the existing system. Advanced-country technology is strongly associated with relatively high wages whereas traditional technology is associated with low incomes. Hence, given the nature of the existing technology, governments may have only limited control over relative prices.
>
> In all these areas the causation runs both ways, with a particular technological system giving rise to mechanisms which are consistent with it, leading to new decisions of a similar kind.[32]

Political leaders in many developing countries do care about the poor and disadvantaged segments of their own populations, but they are trapped by this interlocking alliance of power, privilege, and technology between rich and poor countries, unable to act on their own rhetoric to banish poverty and achieve self-reliance. The more their political economies get entangled with the industrialized countries through tied, concessional aid and through integration of the small modern sector of their economies with the world production system dominated by advanced-country technologies, the more difficult it becomes to pursue different paths of social and economic growth based on alternative technologies better designed to fulfill the unmet needs of the majority of their people.

These political leaders would presumably like to strengthen their political, economic, and technological autonomy and diminish their countries' dependence on other nations but are reluctant to pay the price of the necessary social transformation, which might indeed sweep them from positions of power in their own countries. It is no wonder that once hooked on foreign aid with all its entangling consequences in patterns of economic and social growth and political development, such leaders find it hard to shake the habit, even though foreign aid appears to diminish the capacity of a country to save and thus to build the productive capacity necessary for meeting the minimum material needs of most, if not all, of their people. Subramaniam Swamy, an Indian critic of foreign aid who argued that such aid "provides second-rate irrelevant technology damaging to the Indian economy," posed the issue this way:

> But why should foreign aid sap the will of a nation to save? Because saving is an effort, and consumption is a propensity. Politicians and bureaucrats hate hard decisions, especially those which entail making sacrifices. So, if foreign aid is easily available, why bother to tighten belts—especially if going along with foreign aid means pleasant all-expenses-paid jaunts to Washington, London and Paris? And once the soft option is exercised, then anything goes! In 1976, the then Government negotiated 21 irrigation projects costing Rs. 2,200 crores (approximately US $2.6 billion) with the World Bank. Why would any country require foreign aid for irrigation projects which do not require foreign expertise and exchange?[33]

Indeed, foreign aid, typically tied to suppliers in the donor country or countries, becomes itself a powerful determinant of technological choice by recipient countries. And once a certain range of choices has been made, as Frances Stewart has demonstrated, they tend to shape subsequent choices, thus assuring the donor country of continued "policy leverage" (the polite language in major industrialized-country policy documents for using technology to maintain dominance over poor countries).[34] And thus the unholy alliance of power, privilege, and technology becomes cemented and self-sustaining, inhibiting the development and use of alternative technologies incompatible with or threatening to the dominant interests of this alliance. The critical question then becomes how to break out of this vicious circle of dominance and dependence, with all of its melancholy consequences for the poor and disadvantaged in developing countries.

The problem is compounded by the monopoly position that the industrialized countries have in virtually all areas of advanced technology, in industry, agriculture, health, transportation, communications, and almost any other field in the productive or service sectors. But even more important than their present position of overwhelming superiority is their vastly greater capacity for generating new technology. Indeed, this capacity appears to be further concentrated in a handful of the largest industrialized market economy countries, three of which enjoy a near-monopoly (85 percent) on major industrial innovations made in the twentieth century – namely, the United States (60 percent), the United Kingdom (14 percent) and Germany (11 percent).[35]

8. The Past Experience of Delinkers

In the face of the unholy alliance of power, privilege, and technology that binds most of the Third World to a continued state of dependence on the advanced countries, the outlook is bleak for those developing countries seeking a more autonomous, less dependent role in the international system and an increase in their own problem-solving capabilities so that they can tackle their unmet social needs with their own solutions. Yet an examination of the experience of some developing countries suggests it is not hopeless.

Indeed, the place to begin is probably with historical examples of present-day industrialized countries when they were developing – for example, Denmark, Sweden, Australia, and New Zealand in the nineteenth or early twentieth centuries. According to Dieter Senghaas, who has been engaged in extensive studies of different national experiences with disengagement as a development strategy, a key element in such a strategy is internal income distribution. When distribution is less skewed, internal markets become more significant. These circumstances make it possible for countries like those mentioned above to bear the "learning costs" of meeting the needs of 90–95 percent of the population internally, even though in some cases (e.g., Denmark) foreign trade in very carefully limited areas of the economy was used as an engine of economic growth.[36]

Senghaas argued that in his historical examples, the quest for technological self-reliance does not appear to have loomed large, although the net effect of disengagement was to strengthen the national economy, economically and technologically. He is currently engaged in a series of national case studies from the twentieth century, involving such countries as Albania, North Korea, China, and Cuba. Others would argue that in these and other twentieth century examples of Third World countries, the role of technology assumes a very different and much more significant form because of the growing importance of technology as a factor of production and that therefore, control of its generation and use is a more potent leveraging instrument in shaping economic and political relations within and among countries.[37]

It is worth noting that in all of Senghaas's twentieth century case studies, disassociation or delinking was not sought voluntarily but was imposed externally. Delinking is strong medicine that few national political leaders are likely to propose that their people take on their own initiative. It does involve short-term sacrifices for long-term gains. In fact the "integrationist" school would argue that the same gains can be achieved in other, less costly ways. But when autonomy is correlated with equity as measured by less skewed income distribution, the "integrationist" school has a less than convincing case. Whether disassociation, whether self imposed or externally imposed, will lead to greater equity in the absence of an accompanying social transformation is an issue worthy of more searching study and debate than has yet occurred.

The melancholy fact is that most developing countries are trapped in a vicious circle of dominance and dependence out of which it appears that they can break only with some traumatic experience or dramatic step. "It is like shaking-off a bad habit, or overcoming drug addiction," observed Subramaniam Swamy. "One has to make a clean break with it, and abruptly."[38]

The prime example of disengagement in the Third World is China. Having initiated a far-reaching process of social transformation in the 1950s, China "kicked the habit" of external economic and technological dependence in the 1960s with the abrupt withdrawal of Russian aid. For China, this withdrawal was a traumatic event that triggered an enormous spate of indigenous technological problem-solving of all kinds. The Chinese had no alternative (except for exchanging one set of external dependencies for another, probably more restrictive and onerous, which the Chinese political leadership was unwilling to do) to finding their own solutions to their problems.

It is doubtful that the Chinese leadership, notwithstanding all of the rhetoric during the height of the Cultural Revolution, ever harboured any illusions about staying permanently disengaged from the advanced countries in technology and still achieving the national goal of "catching up" with the advanced countries. It is therefore not surprising that in the mid-1970s China sought to relink itself with the world technological system under its newly announced policy of the "four modernizations"—with modern science and technology (one of the four modernizations) as the foundation for modern agriculture, industry, and defense (the other three).

The vital questions for China are whether it will be able to maintain its more autonomous political and economic posture within the international system while acquiring more and more advanced technology from external sources and whether it will be able to preserve the essential characteristics of the social transformation launched in the 1950s and continued in the 1960s without that transformation being undermined by the "corrupting" impact of bourgeois technology on Chinese social institutions and values. The Chinese leadership obviously thinks it can do both. Time alone will tell.

The Chinese seem to have acquired the necessary base of technological competence to absorb advanced technology rapidly in selected key areas of their modernization plan. At least as important and perhaps more so, they also seem to have developed confidence in their ability to manage complex technology and solve the inevitable problems that arise in adapting it to Chinese working conditions and patterns of productive activity without continuing dependence on the external source of the technology (although Chinese technology-acquisition deals do include provision for short-term training of Chinese technical personnel and for consultant services from the technology source).[39]

Most developing countries are small and all developing countries are smaller than China, leading some to argue that a strategy of selective disengagement is really possible, if at all, only for large countries. It is, of course, generally true that smaller countries, whether industrially developed or developing, have a less autonomous, more dependent position in the international system. This is as true for Sweden as it is for Sri Lanka.

But there are a number of larger Third World countries for which some measure of selective disengagement may be feasible if they really wish to maximize their autonomous capacity for solving their own problems. Some other Third World countries have tried such a strategy to varying degrees and in different ways, and we very much need more analyses, carefully and rigorously done, of their respective experiences so as to be better able to compare their experiences with those of the dependent developers, which are much more numerous and whose experiences are generally better known.[40]

For smaller countries, the concept of collective self-reliance remains a theoretical alternative, although the practical difficulties of implementing such a policy on an economically and politically significant scale are well known. Just as China did not initiate the traumatic step that brought about its disassociation from the international system in the 1960s, these smaller developing countries may find themselves forced, by deteriorating internal conditions and increasingly onerous external terms of technology acquisition, into developing closer economic and technological relationships with their neighbors or like-minded countries elsewhere in the Third World.

There remains the Taiwan anomaly. For if Gustav Ranis's analysis of the Taiwanese and Philippine experiences is anywhere near the mark, Taiwan appears to have achieved a fair measure of technological autonomy and to have met in considerable measure the minimum material needs of most of its people.[41] Yet in no sense can Taiwan be said to have achieved this state by having gone through a period of delinking from the world productive system dominated by advanced-country technology, to which it is closely tied. Perhaps Taiwan is the exception that proves the rule. For every Taiwan, there are dozens of developing countries locked into technological dependence on advanced countries. And it is certainly a moot point as to how many Taiwans (or Koreas or Hong Kongs) the industrialized countries would tolerate. Already protectionist rumblings can be heard from the North.

9. Technological Autonomy and Social Equity

Selective technological and economic disassociation is not a goal in and of itself but one means of working toward increased technological autonomy. Technological autonomy, however it is to be achieved, may be perceived as a national goal (indeed, the broader proposition of maximizing economic and political autonomy and minimizing dependence is generally stated to be a national goal of most sovereign states, rich and poor), but the social consequences of pursuing that goal are certainly what matter most for the mass of the people involved. Here there is considerable evidence supporting a positive correlation between increased technological autonomy and greater social equity.

Social equity as a national goal is here defined as having three basic components:

1. Meeting the minimum needs of most, if not all, of the population.
2. Providing opportunities for productive employment for most, if not all, of the population.

3. Less unequal income distribution, leading to satisfaction of basic social wants (adequate health care, education, etc.) of most, if not all, of the people.

There are many other important nonmaterial needs involving identity and freedom that are to be found in the most forward-looking definitions of basic human needs, but the concept itself is sufficiently fraught with possibilities of misunderstanding, misuse, and political manipulation that it seems preferable not to employ it.[42] At all events the concern here is with trying to define social equity in a politically operational (and therefore, to some extent, measurable) way.

Whether greater technological autonomy will necessarily lead to greater social equity (i.e., the two being causally linked rather than simply positively correlated) may be debatable, but there is evidence to suggest that few developing countries have gone very far in meeting the social equity goals defined above on a sustained basis without considerably increasing their autonomous capacity for technological change.[43]

One thing is clear in the experience of the past thirty years of rich country–poor country relations. Any country that cannot meet the minimum material needs of most of its people most of the time remains highly vulnerable to external economic and political domination and manipulation. The Chinese appear to have learned this early and to have given priority to the steps necessary to achieve that goal. Other countries have tried but have been less successful, partly because their national leadership was unwilling or unable to pay the price, for all the reasons explored in the preceding discussion of the unholy alliance of power, privilege, and technology.

To the degree that greater technological autonomy and social equity are positively correlated, if not necessarily causally linked, and to the extent that a period of selective disengagement helps to facilitate the achievement of greater autonomy, there is another basic complicating factor—namely, the resistance by the large industrialized countries that dominate the world production system to developing countries that seek to disengage themselves from that system. China in the 1950s and 1960s and Chile in the early 1970s are two prime examples. This hostility of dominant actors in the international economic and political system is yet another obstacle to pursuing a path that seeks to maximize technological autonomy and, if the two are positively correlated as some evidence suggests, social equity. This is paradoxical because these dominant actors give considerable rhetorical support to the importance of achieving social equity goals in poor countries. The apparent inconsistency may be explained by a hierarchy of priorities in their external relations—with maintenance of their dominant position in the international system being more important than their espousal of increased social equity in poor countries when the two are in conflict.

10. Formulating a Policy of Technological Autonomy

Whatever may be the social merits of increased technological autonomy, there remains the fundamental question of whether it can be made an operational strategy for national development in any meaningful sense or whether it is simply an intriguing theoretical alternative to engage the attention of scholars isolated from harsh political and economic realities. In short, does the concept have any instrumental value in the real world?

Before addressing that question, it is necessary to state what should be obvious: No student of the role of technology in economic, social, and political change, especially an outsider, can conceivably prescribe development strategies for poor countries. Ultimately, such prescription is nonresponsible, if not irresponsible—precisely the underlying problem with foreign experts in poor countries. Nor is it possible to make generalizations that apply to all, or even many, Third World countries except at a highly abstract level, as each represents a distinctive combination of economic, social, political, and cultural values and institutions.

As Denis Goulet wisely observed in commenting on Guinea-Bissau's development experience, "the lesson of greatest importance is that *the best model of development is the one that any society forges for itself on the anvil of its own specific conditions.*"[44]

What follows, therefore, is intended to be suggestive, rather than prescriptive, and is couched in inevitably abstract terms. At the risk of burdensome repetition, it should be emphasized yet again that technological autonomy as used in this chapter is simply a shorthand way of referring to efforts to increase a country's autonomous capacity for creating, acquiring, adapting, and using technologies necessary to meet that country's economic and social needs. Technological autonomy is not autarky or isolation, but it does mean more autonomous decision making about the choice and use of technology.

Within the framework of this definition, there are at least four characteristics or elements in a policy designed to increase technological autonomy:

1. A conscious and deliberate effort to break away from past dependence on external sources of technology, perhaps through a ban on technology imports (subject to the exceptions noted below). Gradual disengagement appears difficult to achieve because policies leading to greater autonomy get undermined by those economic and political interests within the country that benefit most from maintaining existing links.

2. Selective acquisition from external sources of high-priority technologies in the most disembodied form possible, with emphasis on capital goods technologies for meeting the minimum material needs of the majority of the people and on technologies to increase the value-added component of exports based on local raw materials. A critical factor in making these choices is calculation of the social costs and benefits to different segments of the population, with priority for choices giving greatest benefit to those most deprived and exploited. The extent of disembodiment will depend on the level of technological competence within the country; in general, the higher that level, the more disembodiment possible. ("Disembodiment" refers to "arms-length" technology acquisition without entangling, dependency-perpetuating conditions, such as restrictions on exports, long-term management contracts, or integration within the worldwide productive system of a multinational corporation, ordinarily through control of a local subsidiary of the TNC.)

3. A shifting pattern of international trade that seeks to diversify trading partners to avoid excessive dependence on a small group of countries, especially large industrialized countries, the superior economic strength of which always makes small countries (rich or poor) vulnerable. Diversification implies efforts to strengthen trading links with other Third World countries, although given the extreme imbalance in the existing international economic system, this will be no panacea. This strategy also implies trying to increase the value-added component of primary commodity exports, as suggested in the preceding paragraph, and de-emphasizing foreign trade in an attempt to meet local needs as much as

possible through local productive effort, a step that probably offers greater scope for larger developing countries.

4. Most important of all, vigorous support for local innovations to solve local problems and for revitalization of traditional technologies. This may require opting for less economically "efficient" technologies in the short run by recognizing the longer-term benefits gained from learning to solve one's own problems without continued excessive dependence on outside sources and from the confidence this generates to tackle larger, more complicated problems in the future. It will certainly require building up a human resource base, including *relevant* training and educational facilities. Support for developing local innovations also means commitment to use these innovations on the part of the country's political leadership.

These components of a policy to strengthen technological autonomy are not given in any order of priority beyond emphasizing the basic importance of the last element. The mix clearly will depend on the circumstances of the specific country. A strategy for increasing technological autonomy is in any event but a subset of more comprehensive national economic and social policies, which above all else depend on the determination of political leadership over an extended period of time for their realization. The significance of technology policy within this broader context lies in its leveraging potential as a means of "kicking the habit" of continued dependence on outside solutions to local problems and breaking out of the vicious circle of dominance and lack of autonomy.

11. Some Specific Policy Measures

To translate the foregoing discussion into more concrete terms, here are some specific policy measures that, taken in varying combinations as would be appropriate to the country concerned and *effectively implemented*, would add up to a strategy of strengthening technological autonomy or self-reliance, accompanied by a measure of selective disengagement, as a basis for meeting the minimum material needs of the people and leading to a less skewed income distribution:

1. Restrict sharply technology imports, and prohibit them en-

tirely for nonessential consumer goods. Start with the proposition that a social need can be met by using indigenous skills, knowledge, and equipment and upgrading these, refusing to go outside for a solution unless it has been established as the *only* feasible way of proceeding.

2. When it has been concluded that a technology must be imported, insist on conditions leading to genuine transfer and not mere transplantation, such as:

   a. Training of local technologists and workers in all aspects of the technology, including the process of improving it.

   b. Unpackaging the technology wherever possible so that local scientists and technologists can participate in its implantation, thereby increasing their own skills in creating new productive capacity.

3. Build up local problem-solving and production-creating capacities as rapidly as possible, including enterprise-based research, development, and adaptation units and design engineering and production equipment fabrication skills and experience. Invest most heavily in indigenous research and development in those areas of high national priority where the greatest technology importation is likely to occur, notably capital goods technologies leading to production of goods and services to meet the needs of the mass of people and technologies to increase the value-added component of exported raw materials.

4. Analyze the economic and political costs and benefits, both external and internal, of major technological alternatives to meet critical social needs *before* choices are made. Give preference to those alternatives that will benefit the largest number of people and provide the greatest learning experience.

5. Create "alternative" technology systems that are closely linked to small-scale industrial and agricultural production, giving priority to serving those most in need, and that involve users in problem identification and testing and refinement of solutions.[45]

6. Devise financing mechanisms to build up as rapidly as possible indigenous problem-solving capabilities (e.g., access to industrial production for local problem-solving research, a

tax equal to royalty payments for foreign technology for a
similar purpose, etc.).[46]

7. Diversify trading partners as much as possible, seeking
   especially to expand trade with other developing countries,
   by negotiating new or revised trade promotion agreements.

8. Seek untied foreign assistance that permits and encourages
   maximum use of indigenous or other Third World technical
   services and equipment, even if this means a decrease in the
   overall amount of foreign aid.

Developing countries, and especially poor countries, are in many
ways confronted with a Hobson's choice as they seek to find a way
out of their dependent role in the international system and to meet
the most urgent needs of their people—damned if they do and
damned if they do not. The quest for greater technological
autonomy and the concomitant move toward selective disengagement from the global production system dominated by industrialized countries and their TNCs is a "least worst" strategy. It is certainly no panacea for the problems confronting these countries. But
it is also a strategy that carries with it the ultimate promise of a
more just social order in a more genuinely interdependent international economic and political system rather than the present-day
system, which is characterized by dominance and dependence
among nations and by the exploitation and deprivation of the poor.[47]

12. Implications for Industrialized Countries, the United Nations, and a New World Order

A strategy of strengthening technological autonomy in developing
countries clearly depends primarily on the determination of these
countries and their political leadership to break out of the vicious
circle that now imprisons them. But as the industrialized countries
are part of the problem, they can also be part of the solution. The
first and most important step they can take is to ·pledge
noninterference when poor countries choose to disengage from the
international system and go it alone, in greater or lesser measure.
This seems almost platitudinous advice, were it not for the rather
melancholy postwar record of just the opposite on the part of the
larger industrialized countries.

Industrialized countries should also examine the consequences of
existing public policies that have the net effect of subsidizing the ex-

port of socially inappropriate technologies to poor countries and of perpetuating patterns of technological dependence. Investment guarantees, deferral of tax on foreign income of TNCs, and tied aid that requires use of donor-country problem-solving skills are examples of policies that merit searching scrutiny if the unholy alliance of power, privilege, and technology between North and South is to be effectively challenged.[48]

For United Nations agencies and other multilateral development institutions, a strategy by developing countries to maximize their technological self-reliance would also have important implications. All existing programs of assistance to developing countries by such agencies should be scrutinized to determine the extent to which they perpetuate dependence or contribute to the autonomous capacity for local problem solving. Some encouraging beginnings have been made in recent years by agencies such as the United Nations Development Programme and the World Bank, but much more needs to be done. The 1978 United Nations Conference on Technical Co-operation among Developing Countries and the 1979 UN Conference on Science and Technology for Development may have provided some political impetus for this effort, although it is difficult as yet to perceive.

The task for the United Nations system and other international agencies in the 1980s is to accelerate greatly the trend toward more emphasis on local solutions to local problems. It might be well, in order to give these agencies a specific goal toward which to work, to establish a target for the end of the United Nations Third Development Decade in 1990 of 50 percent use of local (or other Third World) expertise, equipment, and facilities in development assistance projects supported by these agencies.

But if the unholy alliance of power, privilege, and technology is so firmly entrenched, political realism compels us to ask what are the prospects for its being seriously challenged. The answer may lie in the rapidly approaching global equity crisis of the mid-1980s. Sharply deteriorating conditions in poor countries may compel political leaders to pursue radically different development strategies, rather than those based on the trickle-down pattern of the past thirty years, or be swept from positions of power to be replaced by those who have the necessary political determination to follow a path of more equitable, self-reliant development.

In rich countries, political and economic elites, confronted with rampant social pathologies generated by unbridled technological change that is dictated by maximizing short-term economic benefits without regard to social costs, may find themselves similarly confronted with new constellations of opposition. The sheer weight of accumulating evidence that highly centralized technological systems cannot be sustained and that they entail massive social and environmental costs when they fail (e.g., the power blackout in New York in the summer of 1977) will help to tip the political and economic balance.[49]

Any set of policies must be judged in terms of its effectiveness in achieving specified goals over a reasonable period of time. Technological autonomy and its companion strategy of selective disengagement are means to the end of more balanced, equitable patterns of development leading to fulfillment of the minimum materials needs of the people and other goals of increased social equity in poor countries. That end is also an essential precondition for a more just, genuinely interdependent world order.

This is not a formula that guarantees success. The choices are hard, and the more autonomous path demands short-term sacrifices for long-term results. The causal relationship between increased autonomy and increased equity remains obscure. Some past delinkers appear to have succeeded more than others. But after thirty years, no dependent developers (with one or two possible exceptions) have "made it" on a self-sustaining basis, if "making it" is defined as significantly increasing social equity on a self-sustaining basis along the lines set forth in this chapter. As we chart new approaches for the 1980s and beyond, development strategies organized around technological autonomy and disassociation are surely worth considering as one means of confronting the unholy alliance of power, privilege, and technology in the contemporary world.

And gradually, uncertainly, haltingly, this heretofore "unthinkable" approach to development is making its way into the mainstream of international debate on the future of the Third World and North-South relations. The repeated failures or trivial results of major international events at the close of the 1970s to bring about significant Northern concessions to Southern demands (e.g., UNCSTD, UNCTAD V, UNIDO III, and the UN Special Session on the Third Development Decade) are driving home the point to Third World leadership that only through greater reliance on their own efforts will developing countries be able to solve their own problems and achieve a more autonomous, less dependent role in the international system. In an editorial entitled "The Disaggregative 80s," the Society for International Development's journal, *International Development Review*, commented:

In this, our first issue of the 80s, discerning readers will find several signs of a counter-trend towards micronization. . . . This trend. . . could perhaps become one of the key conceptual forces in development during the new decade . . . [and] underlines the need to start from self-development as the pre-requisite for any other kind.[50]

The lead article in this issue is by Dieter Senghaas, who argued the case for disassociation and concluded that:

At the present time it can certainly be assumed that the geopolitical and economic map of the international economy in the year 2000 will be substantially different from what it is today. It would be irresponsible to take it for granted that disassociation will become the predominant development programme of the Third World in the course of time without raising any problems. But that the number of "dissociative cases" will grow is something that can be predicted in the light of the failure of traditional development policy and of the abortive North-South dialogue at international conferences and of the ever more aggravated social conflicts in the Third World.[51]

Notes

This chapter is a revised version of a paper prepared for the International Workshop on Technological Dependence – A Major Hindrance for Autonomous Development, sponsored by the German Society for Peace Research, Bonn, Federal Republic of Germany, 2–5 November 1978.
A part of this chapter (the sections on problems of definition, measurement, and analysis) was prepared while I was on an assignment for the UNCTAD Division of Technology Transfer in Geneva. I am indebted for numerous insights to colleagues in that division, particularly those with whom I collaborated in producing a discussion paper on issues and ideas for action in relation to the 1979 United Nations Conference on Science and Technology for Development as one of UNCTAD's contributions to preparations for UNCSTD (Carlos Contreras, Charles Edquist, Osita C. Eze, Ward Morehouse, B. V. Rangarao, and Mervyn Wijeratne, *Technological Transformation of Developing Countries*, Discussion Paper No. 115 [Lund: Research Policy Program, University of Lund, February 1978]; subsequently published by UNCTAD as *Technological Transformation of the Third World* [UNCTAD/TT/9]).
I am similarly indebted for dialogue and criticism that has helped sharpen my ideas on alternative patterns of North-South technology relations and for specific comments on draft versions of this paper to several colleagues at the Research Policy Institute in Lund and to Claude Alvares, David Dickson,

Dieter Ernst, Arthur Ewing, Denis Goulet, Francisco Sagasti, Dieter Senghaas, and Frances Stewart. Needless to say, none of them is in any way accountable for the ideas presented here, with some of which several of them disagree.

Parts of this chapter were presented at the annual meeting of International Studies Association in Washington in February of 1978, and the entire chapter was discussed at the Bonn Workshop on Technological Dependence in November of 1978. Those occasions also generated useful reactions and criticisms that have helped me to refine my ideas. Finally, I want to thank my colleagues in the UNITAR research team on UNCSTD, Jurg Mahner and Volker Rittberger, for their suggestions. I alone, of course, am responsible for the content of the chapter.

1. For a discussion of science and technology as an issue within the context of United Nations conference politics, see Volker Rittberger, *The New International Order and United Nations Conference Politics: Science and Technology for Development as an Issue Area*, UNITAR Science and Technology Working Paper Series, No. 1 (New York: United Nations Institute for Training and Research, 1978).

2. See, for example, Ad Hoc Planning Group for the National Conference on Science, Technology and Development, Inter-agency Working Group on Technology, *Final Report on Alternative Arrangements for the National Conference on Science, Technology and Development* (Report to Under Secretaries' Committee on North/South Policy) (Washington, D.C.: U.S. Department of State, 6 January 1977). A more recent expression of the same thesis is the following from President Carter's national security adviser, Zbigniew Brzezinski: "This [the emergence of more advanced Third World countries like Brazil and Korea as new centers of rapid industrial growth] means that the older industrial countries have to rely increasingly on technological innovation to maintain their place in the world." ("The World According to Brzezinski" [Interview with James Reston], *New York Times Magazine*, December 31, 1978, p. 10.)

3. Galal A. Amin, "Dependent Development," *Alternatives*, Vol. 2 (1976), pp. 379–403; Dieter Ernst, "Technological Dependence and Development Strategies," Attachment to *Lund Letter on Science, Technology and Basic Human Needs*, No. 3 (December 1977). See also Ashok Parthasarathi, "Self-Reliance as Alternative Strategy for Development," *Alternatives*, Vol. 2 (1976), pp. 365–377, for a report on the Pugwash Symposium on Self-Reliance, held in Dar es Salaam, June 1975; and W. K. Chagula, B. T. Feld, A. Parthasarathi, and P. J. Lavakare, eds., *Pugwash on Self-Reliance* (New Delhi: Pugwash Conferences on Science and World Affairs, 1977), a monograph based on this symposium.

4. J. Barzellato, "Self-Reliance and International Collaboration in Science and Technology," in Chagula et al., *Pugwash on Self-Reliance*, pp. 174–181.

5. See, for example, Francisco R. Sagasti, "Underdevelopment, Science and Technology: The Point of View of the Undeveloped Countries—A Discussion Paper," *Science Studies*, Vol. 3 (1973), pp. 47–59. For the dependency-inducing and -perpetuating and economically and socially distorting consequences of continued technology imports by developing countries, see P. Mohanan Pillai and K. K. Subrahmanian, "Rhetoric and Reality of Technology Transfer," *Social Scientist*, No. 54-55 (January-February 1977), pp. 73–92; and Frances Stewart, *Technology and Underdevelopment* (London: Macmillan, 1977). These issues are also discussed in Ward Morehouse and Jon Sigurdson, "Science, Technology and Poverty: Issues Underlying the 1979 United Nations Conference on Science and Technology for Development," *Bulletin of Atomic Scientists*, Vol. 30, No. 10, December 1977, pp. 21–28.

6. See Jon Sigurdson, *Rural Industrialization in China* (Cambridge: Harvard University Press, 1977); Gustav Ranis, "Appropriate Technology in the Dual Economy: Reflections on Philippine and Taiwanese Experience," paper prepared for the International Economic Association Conference on Economic Choice of Technologies in Developing Countries, Teheran, Iran, September 18–23, 1976. This issue is further discussed in Ward Morehouse and Jon Sigurdson, "Science, Technology and Poverty." China is of course now going through a process of relinking technologically with the more advanced industrial nations, as is discussed below.

7. See various papers presented at the International Workshop on Technological Dependence: A Major Hindrance for Autonomous Development (Bonn, 2–5 November 1978), for example, Seifeddine Bennaceur and François Geze, "New Forms of Technological Dependence in Developing Countries Inherent in the Worldwide Organization of the Electronics Industry." For a description of the current global distribution of R&D activity, see another paper for the same workshop, Jan Annerstedt, "Technological Dependence: A Permanent Phenomenon of World Inequality?" These papers are being published in a volume edited by Dieter Ernst, *The New International Division of Labor, Technology and Underdevelopment: The Consequences For the Third World* (Frankfurt: Campus Verlag, 1980).

8. This question is explored further in Carlos Contreras, Charles Edquist, Osita C. Eze, Ward Morehouse, B. V. Rangarao, and Mervyn Wijeratne, *Technological Transformation of Developing Countries: Some Issues for Discussion and Preliminary Ideas for Action at the National and International Levels in the 1980s* (paper based on informal consultations organized by UNCTAD, Geneva, January 1978), Discussion Paper No. 115 (Lund, Research Policy Program, University of Lund, February 1978), especially pp. 1–4. (This paper has also been published as UNCTAD, *Technological Transformation of the Third World: Issues for Action in the 1980s* (Report on Informal Consultations Organized by the UNCTAD Secretariat, Geneva, January 1978) (Geneva: United Nations, 1978) (UNCTAD/TT/9). One of the most articulate statements on technological autonomy in the Third

World is by Francisco R. Sagasti, "Technological Self-Reliance and Cooperation among Third World Countries," *World Development*, Vol. 4 (1976), pp. 939–946. See also Chagula et al., *Pugwash on Self-Reliance*, which contains this paper by Sagasti and contributions by others on the same theme.

9. On unbundling external technology acquisitions, see UNCTAD Secretariat, *Handbook on the Acquisition of Technology by Developing Countries* (New York: United Nations Conference on Trade and Development, 1978) (E.78.II.D.15).

10. The importance of revitalizing traditional technologies is stressed by Francisco Sagasti in "Endogenization of the Scientific Revolution," *Human Futures*, Summer 1978.

11. See United Nations Conference on Trade and Development (UNCTAD), *Guidelines for the Study of the Transfer of Technology to Developing Countries* (New York: United Nations, 1972) (TD/B/AC.II/9).

12. Francisco R. Sagasti, "Towards a New Approach for Scientific and Technological Planning," *Social Science Information*, Vol. 12, no. 2 (1973), pp. 67–95.

13. Francisco R. Sagasti, letter to the author, October 23, 1978.

14. UNCTAD, *Case Studies in Transfer of Technology: Policies for Transfer and Development of Technology in Pre-war Japan* (1807–1937 period), Study by the UNCTAD Secretariat, Geneva: UNCTAD, 1978 (TD/B/C.6/. . . .in press); UNCTAD, *Transfer of Technology: Action to Strengthen the Technological Capacity of Developing Countries — Policies and Institutions* (Report by the UNCTAD Secretariat, 4th Session, Nairobi, May 1976) (Geneva: UNCTAD, 1976) (TD 190/Supp. 1).

15. See P. Mohanan Pillai and K. K. Subrahmanian, "Rhetoric and Reality of Technology Transfer," and the study by Alejandro Nadal of Mexican technology policies of the Science and Technology Policy Instruments (STPI) Project (Spanish edition published by El Colegio de Mexico, Mexico City; English edition forthcoming). See also Lynn K. Mytelka, "Regulating Foreign Investment and Technology Transfer in the Andean Group," *Journal of Peace Research*, Vol. 14, no. 2 (1977), pp. 165–182; and Francisco R. Sagasti, *Science and Technology for Development: Main Comparative Report of the STPI Project* (Ottawa: International Development Research Centre, 1978).

16. William Arthur Lewis, *Theory of Economic Growth* (Homewood, Ill.: R. D. Irwin, 1955), especially Chapter 4 on "Knowledge," pp. 164–201. David Dickson has rightly pointed out that it is the control of the creation, acquisition, and use of knowledge, rather than knowledge itself, that is a source of power in the modern world. (David Dickson, presentation to NGO Forum on Science and Technology for Development, New York, September 1978).

17. V. M. Dandakar and N. Rath, *Poverty in India* (Poona: Indian School of Political Economy, 1971). For an extended discussion of the methodological problems involved in determining income distribution in

India, see Pranab K. Bardhan, "The Pattern of Income Distribution in India: A Review," *Sankhya: The Indian Journal of Statistics,* Vol. 36, Series C, Parts 2 and 4 (1974), pp. 103–138. Estimates of poverty in India vary; Bardhan, for example, estimated that in 1968–1969, 54 percent of the rural population was below his definition of the poverty line of Rs. 15 per capita per month at 1960–1961 prices (about $3 at then prevailing exchange rates). This discussion of the global equity crisis is adapted from Ward Morehouse, *Science, Technology and the Global Equity Crisis: New Directions for United States Policy,* Occasional Paper No. 16 (Muscatine, Iowa: Stanley Foundation, May 1978).

18. K. N. Raj, "The Economic Situation," *Economic and Political Weekly* (Bombay), July 3, 1976, p. 995. The relationship of technology to poverty and socioeconomic equity is further discussed in Ward Morehouse and Jon Sigurdson, "Science, Technology and Poverty."

19. James H. Weaver, Kenneth P. Jameson, and Richard N. Blue, "A Critical Analysis of Approaches to Growth and Equity," paper prepared for International Studies Association Annual Meeting, March 17-20, 1977, St. Louis. See also Hollis Chenery et al., *Redistribution With Growth* (Oxford: Oxford University Press, 1974), and Irma Adelman and Cynthia T. Morris, *Economic Growth and Social Equity in Developing Countries* (Stanford, Calif.: Stanford University Press, 1973).

20. For a discussion of varying viewpoints on this issue, see Weaver et al., "A Critical Analysis of Approaches to Growth and Equity."

21. Robert S. McNamara, Address to the Board of Governors, Washington, D.C., World Bank, September 26, 1977, p. 7.

22. Sarwar Lateef, *Economic Growth in India and China, 1950-1980,* EIU Special Report No. 30 (London: Economist Intelligence Unit, 1976). There is, of course, more to the respective Indian and Chinese experiences with industrialization than this simple statement may suggest. Thus, India's exports of nontraditional manufactures have zoomed upward in the last five years, but for the past decade, employment in the modern private industrial sector has remained almost static, while unemployment, underemployment, and employment at very marginal rates of productivity have increased substantially.

23. McNamara, Address to the Board of Governors, World Bank, p. 9.

24. See, for example, Amon J. Nsekela, "The World Bank and the New International Economic Order," *Development Dialogue,* No. 1 (1977), pp. 75-84.

25. See the October 6, 1977, letter from President Jimmy Carter to Congressman Clarence Long, stating that Carter would instruct the U.S. representatives to the multilateral development banks to vote against financing of projects that would increase production in developing countries of palm oil and several other commodities. Carter's letter was the price the administration had to pay for removing from the congressional appropriation of U.S. funds for these banks a categorical prohibition against using U.S.

contributions to the banks to increase production of those commodities asserted by U.S. producers to be hurting the market for U.S. production.

26. Carlos F. Diaz-Alejandro, "Delinking North and South: Unshackled or Unhinged," in C. F. Diaz-Alejandro et al., *Rich and Poor Nations in the World Economy* (New York: McGraw-Hill for the Council on Foreign Relations 1980s Project, 1978).

27. On this unholy alliance in relation to UNCSTD, see A.K.N. Reddy, "Science v. Deprived Humanity," *New Scientist,* 26 October 1978.

28. David Dickson has argued, in a letter to the author (31 October 1978), that industrialized-country market economy

> capital, in order to expand, requires a certain set of political relationships (a docile labour force as a source of wage-labour, etc.) and . . . technological policies are manifestations not only of the economic relationships between the industrialized and the developing nations, but also of these political relationships, which are not merely broad relations of the domination/subordination type in the more traditional sense but, with the growth of the internationally-integrated economy, reach down to the discipline and control needed in the heart of the labour process itself.

29. Stewart, *Technology and Underdevelopment*, p. 277.

30. It should be emphasized that alternative technologies do not necessarily mean "intermediate," "soft," "village," or "small-scale" technologies but also include large-scale, scientifically sophisticated, or complex technologies, *albeit* typically with different factor endowments or satisfying social needs in different ways than technologies now being used in industrialized countries.

31. See, for example, G. S. Aurora and Ward Morehouse, "The Dilemma of Technological Choice in India: The Case of the Small Tractor," *Minerva,* Vol. 12, no. 4 (October 1974), pp. 433–458. Also relevant is the critical study of Mexico's science and technology policies by Alejandro Nadal and other national case studies generated by the Science and Technology Policy Instruments Project (STPI) mentioned in note 15.

32. Stewart, *Technology and Underdevelopment*, pp. 276–277.

33. Subramaniam Swamy, "Addiction to Foreign Aid," *India Today,* May 1–15, 1978, p. 61. Swamy's figure for World Bank aid to India is exaggerated. The enervating and dependency-inducing effects of foreign aid have been examined in the work of Denis Goulet (e.g., *The Uncertain Promise: Value Conflicts in Technology Transfer* [New York: IODC/North America, 1977]) and Theresa Hayter (e.g., *Aid as Imperialism* [Harmondsworth: Pelican, 1971]).

34. See U.S. State Department Ad Hoc Planning Group for the National Conference on Science, Technology, and Development, *Final Report on Alternative Arrangements for the National Conference on Science, Technology and Development.*

35. Organisation for Economic Co-operation and Development, *Gaps in*

Technology: Analytical Report (Paris: OECD, 1970), as cited in Alan Maislisch, "Trends in the Supply of Technology," unpublished paper, May 1975.

36. See, for example, Dieter Senghaas, "Disassociation as a Development Rationale," paper prepared for International Workshop on Technological Dependence, Bonn, 2-5 November 1978, and "Alternative Entwicklungsstrategien Dissoziation und Autozentrierte Entwicklung – Eine Entwicklungspolitische Alternative für die Dritte Welt," *Internationale Entwicklung*, No. 3 (1978), pp. 27-46.

37. Senghaas's comparison by historical analogy of the role of technological autonomy in nineteenth and early twentieth century First World and late twentieth century Third World national development experiences was sharply challenged by other participants in the International Workshop on Technological Dependence (Bonn, 2-5 November 1978).

38. Swamy, "Addiction to Foreign Aid."

39. For a fuller discussion of the Chinese experience, see the various works of Jon Sigurdson, especially his *Technology and Science in the People's Republic of China: An Introduction* (New York: Pergamon Press, 1980), and the version of this chapter being published as part of the proceedings of the Bonn Workshop on Technological Dependence, for which it was originally prepared.

40. One such analysis is that by Denis Goulet, *Looking at Guinea-Bissau: A New Nation's Development Strategy*, ODC Occasional Paper No. 9 (Washington: Overseas Development Council, 1978).

41. Ranis, "Appropriate Technology in the Dual Economy."

42. See, for example, the list of basic human needs by Johan Galtung, Director of the Goals, Processes and Indicators of Development Project (Geneva: United Nations University/UNITAR, 1977).

43. For a brief discussion of the national development experiences of such countries as Taiwan, Burma, Kuwait, and Sri Lanka to sustain this generalization, see the version of this chapter being published as part of the proceedings of the Bonn Workshop on Technological Dependence, for which it was originally prepared.

44. Goulet, *Uncertain Promise*, p. 52 (emphasis in the original).

45. Some of the characteristics of "alternative" or local technology systems are described in Contreras et al., *Technological Transformation of Developing Countries*, especially Annex II, p. 27.

46. A number of financing mechanisms have been set forth by Francisco R. Sagasti in "Towards an Endogenous Scientific and Technological Development for the Third World," in Ward Morehouse, ed., *A New Science and Technology Order* (New Brunswick, N.J.: Transaction Books, 1979).

47. I have spelled out the nature of this Hobson's choice in a background paper for the Jamaica Symposium on Mobilizing Technology for Development, "Disengagement and Third World Collaboration: Neglected Options in Mobilizing Technology for Development" (Washington: International In-

stitute for Environment and Development, 1978); an abbreviated version of this paper was subsequently published in Jairam Ramesh and Charles Weiss, Jr., *Mobilizing Technology for World Development* (New York: Praeger Publishers, 1979). See also Peter O'Brien's paper for the International Workshop on Technological Dependence (Bonn, 2–5 November 1978), "The Eternal Triangle: Another Look at the Relationships Between Development, Industrialization and Technology," published in Ernst, *The New International Division of Labor*, pp. 503–528, for identification of some of the transitional steps necessary to achieve this kind of new social order.

48. I have attempted to spell out a set of such policy initiatives for the United States in Morehouse, *Science, Technology, and the Global Equity Crisis*.

49. For a thoughtful analysis of the present predicament of industrialized societies and a coherent statement of alternative future directions for social and technological change, see James Robertson, *The Sane Alternative* (London: Published by the Author, 1978).

50. Editorial for issue on "Development from Within," *International Development Review*, 1980/1, p. 2.

51. Dieter Senghaas, "The Case for Autarchy," *International Development Review*, 1980/1, p. 10.

3
Third World Cooperation in Science and Technology for Development

David W. Chu
Ward Morehouse

1. The Global Economic and Political Context for Third World Cooperation: The Changing Role of Science and Technology in the International System

Global interdependence is a fact of late twentieth century life. We are all riders on Spaceship Earth—but alas, while a small proportion of humanity travels first class, the rest are down in steerage. Melancholy evidence accumulates to show that the "benefits" of international cooperation for development flow unevenly both within and among countries. This is hardly surprising when the rules of the game have been set by the more powerful actors in the international economic and political system.

Although all the riders on the global spaceship share an ultimate concern for improving the human condition, it is manifestly not in the short-term interests of powerful actors in the international system to surrender willingly their superior capacity for generating and using technology to produce goods and provide services. The technology they do generate, furthermore, is designed to serve their needs and is largely irrelevant to the needs of poor countries and people. Indeed, one may make the case that behind the facade of international cooperation in science and technology for development lies a harsher reality: the preoccupation of some of these more powerful governmental and corporate actors with using their superiority in science and technology to hook the poor countries in a dependent relationship and thus to maintain dominance and control.

The emerging importance of technology as an instrument of domination and control in the international system was well expressed recently by a senior foreign policy official of one of the world's leading scientific and technological superpowers when he observed that the emergence of more advanced Third World countries as new centers of rapid industrial growth "means that the older industrialized countries have to rely increasingly on technological innovation to maintain their place in the world."[1] The relative decline in overt forms of military and economic power of the major industrialized countries is leading them to turn more and more to technology as a means of maintaining their position in the international economic and political system.

This trend is accompanied by several others, all pointing to science and technology as increasingly significant factors in relations between nations in the 1980s and beyond. Among these trends is the persisting, if not widening, gap between North and South in the capacity for technological advance across the full range of economic, social, and military activities in which technological change plays a significant role, resulting in increasing technological dependence of the South on the North.

Notwithstanding this persisting gap and growing dependence, the past thirty years have seen the establishment of beachheads of scientific and technological competence in a small but growing number of Third World countries. These beachheads encompass a variety of sectors of productive activity, basic research, and social services, including high-technology fields such as electronics, space, and computers.

Another trend affecting the global context for science and technology in the Third World is the rapid deceleration in rates of economic growth in industrialized societies, beginning in the mid-1970s. Because Third World economies are still tied to industrialized-country economies, there has been a corresponding deceleration in growth rates in developing countries. These developments are further complicated by emerging resource scarcities in such basic areas of human need as food and energy—resource scarcities that apparently will become worse, with growing world population and demand, long before they improve.

2. Linkages of Power, Privilege, and Technology in the Global Economic and Political System

Underlying the irrelevance of most technological change to the unmet social needs of the majority of the Third World population are the linkages between the rich countries that collectively generate their own technology and that 10–20 percent of the population of poor countries that have adopted rich-country consumption patterns. Such groups feel little compulsion to break their links to technological development in the advanced countries through investment, trade, and the other economic and political ties that are the carriers of technology from North to South.[2]

Notwithstanding these linkages, most Third World political leaders express deep concern for the poor of their countries. They really want to eradicate poverty from their lands. But the sad fact is that their maneuverability is limited. The commitment of Northern elites to fighting poverty in the Third World is even more dubious. Their compulsions are less intense and the domestic political costs are likely to outweigh by far the marginal economic and social benefits to their own constituents. Given these circumstances, bold talk about mounting global wars on poverty lacks credibility.[3]

Many developing countries are placing major emphasis today on the development of their own technological capacity in order to enhance their economic and political autonomy and to diminish their dependence on industrialized countries. They argue that this enhanced autonomous capacity for technological change is an essential precondition to their ability to solve their economic and social problems, including meeting the minimum material needs of their peoples. Whether greater autonomy will necessarily lead to greater equity may be arguable, but there is some evidence to suggest that few developing countries have gone very far in meeting the minimum needs of most of their people on a sustained basis without substantially increased autonomous capacity for technological change.

Industrialized countries should hardly find surprising this concern with technology as an instrument of national power. The era of "big" science and technology after all began during the Second World War, when science and technology were perceived by political leaders in those countries as a means of strengthening their military advantage over their adversaries. This interest in technology as an instrument of national power has dominated massive public investments in R&D in industrialized countries ever since, even though in more recent years more attention has been paid to social considerations and the quality of the physical environment.

3. An Orwellian Scenario for the 1980s

Nonetheless, because of the strong linkages of power, privilege, and technology and because of the importance of a modern industrial sector as a foundation for national political power, the pursuit of a trickle-down development strategy based on maintaining advanced-country technology links remains strongly tempting to Third World political leaders and is actively encouraged by some Northern elements that stand to benefit from keeping or strengthening these links. The temptation is compounded by the bleak character of the alternatives, but those in the Third World opting for this strategy would do well to ponder the implications for their fragile modern industrial economies of emerging trends in Western technology.

Let us assume we are now in the mid-1980s. The most industrialized Southern market economy countries, which are generally regarded as the more advanced within the Third World in terms of industrial and allied technological capacity, are still characterized by a sharp inequality of income and widespread rural unemployment and underemployment, but with a substantial modern industrialized sector in their dual economies. The leading industrialized countries, responding to continuing pressure to increase productivity of labour because of rising compensation levels at home, have automated much of their production technology to the point of eliminating relatively labour-intensive manufacturing operations in developing countries and moving toward an increasingly self-contained industrial system among industrially developed countries. This trend may be accelerated by rising labour costs within such developing countries and their growing assertiveness on international economic and political issues.

These more advanced developing countries will have become increasingly integrated technologically with the industrialized countries, especially through transnational corporations (TNCs), which seek to rationalize production on a global scale to minimize costs and maximize profits. However, they will lack the critical capacity to innovate because they have never had to try and because that capacity has been a carefully guarded monopoly of the TNCs and their home-country governments.[4] In the 1980s the newly industrializing countries (NICs) of the Third World, if they persist in their present policies, may find themselves with an installed production capacity that will virtually overnight become obsolete. They will have only limited internal markets to use what this capacity might produce because of high income inequality and mass poverty. And they will not have the ability to adapt quickly in response to the cataclysmic changes that have been externally determined without reference to their own needs. In short, they will be left high and dry.

4. Autonomous Development as an Alternative Strategy

But consider a possible alternative scenario, involving those developing countries that in the 1960s and 1970s recognized the critical importance of strengthening their autonomous capacity for choice and the use of technology to meet their needs. They will have placed heavy emphasis on creating local technology systems to improve productivity in the rural areas to meet minimum needs. In their industrialization efforts, they will have concentrated on processing their own raw materials, which they control, and avoided or de-emphasized a role in the global production system dominated by industrialized country–based TNCs, in which they as developing countries with relatively low-cost labour were expected to perform those labour-intensive phases of the production process for consumer and intermediate goods primarily destined for industrialized-country markets.

These same countries will have concentrated on buying technology from abroad in the most disembodied form possible, even though it usually costs more initially, simultaneously investing more heavily in their own research and development in precisely those areas where they were buying foreign technology so as to be able to improve upon what they acquired elsewhere. When technology was acquired from abroad, the highest priority will have been given to those technologies that would enable the country better to meet the minimum needs of the majority of its people. At the same time these countries will have sought to strengthen their trade with other developing countries, including participation in Third World multinational enterprises that pool capital and technological skills available from within the Third World, in order not to be any more vulnerable than necessary to changes in major industrialized-country economies.

These countries, although remaining a part of the world market system, will clearly be in a markedly different situation a decade hence than the NICs, assuming that the latter continue to follow a pattern of technologically dependent economic growth. As long as the capacity for technological innovation remains concentrated in the industrialized countries, developing countries will be vulnerable to technological choices made elsewhere.

Industrialized-country choices are unlikely to be made in order to protect the interests of developing countries, especially where vital interests within industrialized countries are involved. On the contrary, the industrialized countries will make choices in technology, and more broadly in industrial, agricultural, and energy policy, to minimize their dependence on others as long as the individual nation-state remains the key actor in the international political and economic system. Developing countries would be well advised to follow the same strategy.[5]

5. Disengagement and the Evolution of the International Economic Order

Measured disengagement from the international economic and political system dominated by the industrialized countries may seem to many an unthinkable posture for developing countries to adopt. There are, as we have noted, certainly powerful forces operating against such disengagement. But no less an authority on the economic development of the Third World than W. Arthur Lewis has reminded us recently that current relationships are not among the permanent ordinances of nature.[6] Indeed, disengagement as a national development strategy has considerable historical precedents. Almost all of the industrialized countries went through periods of relative disengagement from the international system at different periods (though admittedly each case must be treated separately because the cause for such disengagement in the past and its relationship to development are not always clear).

When Third World countries switch from exporting primary products to exporting manufactures to the industrialized countries, as they have been doing increasingly in recent years (especially the larger and more advanced developing countries), they exchange one dependence for another. To be sure, the potential scope for economic interaction is much wider. There is, after all, a limit to the amount of tea, cocoa, or coffee that industrialized countries will buy. But with manufactured exports from the Third World standing at only 8 percent of world trade in manufactured goods in 1975, potentially unlimited growth would appear to be available in this area of international trade to developing countries within the next decade or so.

World trade in manufactured goods has been growing by about 10 percent yearly. So have exports of manufactured goods from developing countries. If this rate of growth continued, developing countries would merely be holding a constant proportion of world trade. In the short term, this would not seem to present either North or South with overwhelming problems.

However, it is unlikely that world trade in manufactured goods will grow indefinitely at 10 percent per annum, when global production of manufactured goods grows at only 5-6 percent yearly. Indeed, the deceleration in growth rates in industrialized countries characteristic of the middle and late 1970s is a harbinger of things to come. As the growth rate falls, Third World countries will need an increasing share of world trade in manufactured goods, and this is bound to face increasing resistance. As Lewis noted:

> The fact is that the LDCs should not have to be producing primarily for developed country markets. In the first place, they could trade more with each other, and be less dependent on the developed coun-

tries for trade. The LDCs have within themselves all that is required for growth. They have surpluses of fuel and of the principal minerals. They have enough land to feed themselves, if they cultivate it properly. They are capable of learning the skills of manufacturing, and of saving the capital required for modernization. Their development does not in the long run depend on the existence of the developed countries, and *their potential for growth would be unaffected even if all the developed countries were to sink under the sea.*[7]

Lewis goes on to observe that he does not recommend such total and complete disengagement of the Third World from the industrialized countries. He is enough of a realist and a student of history to know that the more advanced and rapidly industrializing developing countries will continue to look to rich-country markets, to the extent that these markets are open to them, for their growing exports, and not to the tiny markets of other impoverished Third World economies. That is after all what happened in the recent past. Thus, Germany became a major world trading nation in the 1880s by flooding the British market; the United States did the same thing at the end of the nineteenth century in Europe.

But history does not often repeat itself exactly, and although international trade became an engine of growth in the nineteenth century, this is not its proper role today, according to Lewis:

> The engine of growth should be technological change, with international trade serving as lubricating oil and not as fuel. The gateway to technological change is through agricultural and industrial revolutions, which are mutually dependent. International trade cannot substitute for technological change, so those who depend on it as their major hope are doomed to frustration. If we can make this domestic change, we shall automatically have a new international economic order.[8]

Given the extreme maldistribution of the world's problem-solving capacity through science and technology (with 95 percent of global expenditure on research and development concentrated in a handful of larger industrialized countries) and with the terms of access to much of the most advanced technology difficult, Third World countries have no realistic alternative but to move forward as rapidly as they can in acquiring their own problem-solving capacities. This is, however, a slow and complicated process. In the meantime, unmet social and economic needs in these countries increase and in some cases are exacerbated by dependence on industrialized-country technology inappropriate for meeting these needs.

An imitative industrial strategy is not likely to lead to meeting the minimum material needs of most of the people. Few countries, even those that claim to control their imports of technology, have been able to avoid the socially distorting effects of a pattern of economic growth that is based on foreign technology (or even an indigenously developed derivative technology) and is geared primarily toward meeting the needs of the largely urbanized middle classes.

The small modern industrial sector in most Third World economies has thus far succeeded in soaking up scarce capital through the importation of advanced-country technology while doing virtually nothing for one of the most critical problems facing these countries—unemployment, underemployment, and employment at very marginal rates of productivity. Lack of productive employment is only one part of the equity crisis in the Third World that will reach arresting proportions by the mid-1980s. Sharply increasing income inequality is another, and acute material deprivation is yet another.

The changing character of science and technology in the international system, the growing equity crisis, the gap between North and South in problem-solving capabilities, technology as an ever more important source of national power—all add up to a powerful set of compulsions for Third World countries in the first instance to strengthen their national capacity for using technology to solve their problems, and in the second instance, to seek greater Third World cooperation, not as a substitute for international cooperation and North-South flows of science and technology, but as a counterweight in an effort to achieve a better balance in their own national efforts.

We may in fact be confronted with a paradox. The path toward a more just world order may be more certain in the long run if poorer countries disengage themselves in varying degrees in the short run from an international system in which they are now weak and dependent actors with little capacity to change the rules of that system to their advantage. Disengagement from the North will serve as a major force in stimulating South-South cooperation in science and technology.

6. Patterns of Third World Cooperation: Trends and Cases

The imperatives for Third World cooperation in science and technology have been recognized by an increasing number of Third

World countries over the past several years. One consequence of this realization was the August 1978 Conference on Technical Cooperation Among Developing Countries, discussed in the section on UN conferences below. Yet such cooperation is hardly new and has already taken many forms.

Cooperation in science and technology among developing countries has occurred as a result of some perceived need and/or advantage. The impetus for such a need or advantage may come from the marketplace, from government sources, or from multilateral agencies, such as those in the UN system. In most cases cooperation among developing countries occurs bilaterally, especially when cooperation is between individual firms or between governments. Only when the nature of the cooperation involves a problem, commodity, or service shared by a number of countries, or when a regional or super-regional authority (e.g., the United Nations) is involved, does the cooperation become multilateral or interregional in character.

Of the three forms of cooperation in science and technology for development discussed below, cooperation among business enterprises occurs with their counterparts in other developing countries through the normal market incentives for profits. Such cooperation typically involves the buying and selling of technology in various forms (turnkey plants, machinery, consultative services, joint ventures, etc.)

Third World governments have a number of reasons for cooperating with other developing-country governments in science and technology. Where government-owned industries are involved, the incentives for cooperation may be financial, as they are in the private sector. In other cases a government may decide to initiate cooperative ventures in industry, agriculture, or other areas of economic activity not yet fully developed in its own economy, thereby requiring help and cooperation from other countries or international agencies. Finally, a developing country may be motivated to cooperate with other developing countries by other kinds of concerns, often involving a public good, such as national security or the control of natural resources.

The third form of cooperation involves the intervention of international agencies like the United Nations to promote cooperation among governments and private firms in developing countries. Such agencies have often been helpful in providing financial support and expertise from other developing countries and in encouraging regional and interregional cooperation in science and technology within the Third World.

The forms of cooperation that have already occurred do not, of course, exhaust the range of possibilities open to developing countries. After examining actual cases in this section, we shall look at various obstacles to and opportunities for Third World cooperation in the next two sections and at ideas and proposals for action in the section following those.

A. Enterprise-to-Enterprise Links

In his paper "Developing Countries as Exporters of Industrial Technology," Sanjaya Lall listed five kinds of Third World cooperation that occur in the market place. These five kinds of cooperation primarily take the form of an exchange of technology between two countries (or more specifically between firms within those countries). These include the setting up of entire production systems (turnkey projects), engineering consultancy, licensing of patents and managerial and technical services, direct investment, and training schemes.[9]

The principal exporters of entire industrial plants in the Third World are firms in India, Argentina, Brazil, and Mexico. The types of industries represented in such kinds of exports include textiles, electrical generation, fertilizer, machine tools, meat refrigeration, fruit processing, and steel. The transfer of turnkey plants of these sorts is usually a sophisticated operation and requires an indigenous technology and production base that only the more advanced of the developing countries have. Furthermore, in countries like India where many of the industries are owned by the government, the interface between private and public exchange of technology is less distinct than in countries where this condition does not exist. In such cases as India, many enterprise-to-enterprise transfers of technology are in fact government-to-enterprise transfers.[10]

The transfer of engineering consultancy services in the Third World often occurs in areas involving relatively complex technology, such as power generation, mining, oil, petrochemicals, chemicals, paper, and steel. Again, it is countries like India, Mexico, Brazil, and Argentina that possess the greatest number of consultancy enterprises.[11]

Licensing of patents and trademarks and the transfer of managerial and technical services seem to occur infrequently among developing countries, perhaps because of the dominance of the industrialized countries in these fields. There are exceptions, however. Managerial services are exported by Indian hotel chains, and Indian Railways has provided technicians to some African countries.[12]

Direct investment of Third World firms in other Third World countries is becoming increasingly common. In Latin America, these Third World transnational firms specialize mostly in consumer products requiring "low to medium" level technology, such as electrical, food, and metal products. Argentina, Mexico, Brazil, and Peru lead the way in terms of the number of affiliates abroad. In Asia, Hong Kong has established affiliates in textiles, and Singapore and Korea have transnational firms in such industries as electronics, engineering works, and food products. Indian industries with overseas subsidiaries or affiliates include textiles, flour mills, steel products, pharmaceuticals, and many others.[13]

In terms of training programs set up in developing countries by private enterprise in other developing countries, India again seems to lead the way. Its firms have sponsored technical training institutes in Singapore, Iran, Guyana, and Malaysia. Like licensing and management techniques, however, this form of technology transfer is still dominated by the industrialized countries.[14]

B. Government Cooperation

Governments cooperate in exchanging technology with other countries under any of the following conditions: The private sector in the industry of concern is not sufficiently developed; there are inadequate financial incentives or profit opportunities for private enterprises; a public good is involved; or the public and private sectors are already intimately connected. Although, as we have seen, the exchange of technology by private enterprise originates in only a handful of the relatively more advanced developing countries, this need not be the case in order for governments to cooperate with one another. This is possible because the conditions for the science and technology infrastructure necessary to maintain a relatively advanced enterprise system, as in India or some of the Latin American countries, need not be the same for governments, which command far greater resources than individual firms, to engage in technology exchange with other countries. As a result, some of the least developed countries have been able to cooperate in science and technology through their governments even though the conditions may not exist for private enterprise to engage in technology transfer in a similar way.

Numerous forms of cooperation in science and technology have occurred in Latin America, Africa, and Asia. Such cooperation is usually motivated by a desire to solve a shared problem. In Latin America, the Central American republics and Panama have joined in some eighteen projects to solve various energy and environmental problems that affect them. These projects include the integration and development of telecommunications, technical consultancy in the fertilizer industry, a hydro-meteorological project, and prevention and damage-limitation systems for hurricanes and floods.[15] For example, the Salto Grande Hydroelectric Project has drawn technical resources from Argentina and Uruguay as well as from other developing countries in order to formulate and implement strategies and programs for environmental protection and for occupational health and human settlements. Technical cooperation agreements have also been made between Argentina, Brazil, and Paraguay to implement joint action on two large hydroelectric schemes that will be constructed at Itaipu and Yacireta.[16]

In the problem area of health care, Egypt is providing Sudan with teachers in medicine, dentistry, and pharmaceuticals, and an institute for liver diseases affiliated to the one in Cairo is to be established in Khartoum. Similarly, Cuba and Mexico have signed an agreement for the exchange of information and of experts and training in the health field, just as Peru and Colombia have jointly developed programs for the training of personnel and the development of national programs for food hygiene.[17] In Africa, with the assistance of the World Health Organization (WHO), Guinea, Mozambique, Senegal, Tanzania, and Zambia are cooperating with Cape Verde and Guinea-Bissau and with other newly independent Portuguese-speaking countries in Africa in the reconstruction of their health services.[18]

Cooperation may result from the need to solve a technological problem that affects a particular commodity shared by countries in a region. For example, in Central America, Nicaragua, Guatemala, Mexico, Honduras, El Salvador, Panama, and Costa Rica have agreed to contribute $12 million to a fund to combat coffee leaf rust disease. Similarly, the Andean countries of Bolivia, Ecuador, Colombia, Peru, and Venezuela formed in 1977 a Technical Committee of Andean Coffee Rust Specialists in order to study methods to prevent and combat this disease, which destroys a vital commodity.[19]

The member countries of the Association of Natural Rubber Producing Countries (ANRPC) in Asia have cooperated in studying the demand for and supply of natural rubber and in determining the maximum production capacity of natural rubber in each of the member countries. There has also been cooperation in field surveys among coconut-producing countries and a study of socioeconomic measures needed to improve the productivity of pepper holdings in the Pepper Community member countries.[20]

In the Asia-Pacific region a network is being established by India, Indonesia, Iran, the Republic of Korea, Pakistan, the Philippines, Sri Lanka, and Thailand in order to evaluate the actual and potential range of farm mechanization systems. This includes identifying bottlenecks in the mechanization of farming methods, promoting local manufacture of appropriate agricultural machinery, and acting as a clearing house for the exchange of information on farm mechanization programs.[21]

For the production and processing of sugar, the Group of Latin American Caribbean Sugar Exporting Countries has promoted technical cooperation with respect to that commodity. It issues information on the most advanced sugar technology and encourages the interchange of technical personnel through training programs.[22]

In June 1977, officials from Botswana, Kenya, Malawi, Mauritius, Swaziland, Tanzania, and Zambia came to Lesotho to study Lesotho's cooperative and self-help housing project. Designed to provide adequate low-cost housing, this project is an example of how even one of the least developed countries can offer its experiences in a particular technology for the benefit of other developing countries.

As a final example of governmental cooperation in science and technology, a number of developing countries have made available their technical facilities to other developing countries. Malaysia and the Republic of Korea have offered the facilities of their isotopic age-dating laboratories to neighbouring countries in the region. The Philippines is prepared to make available to other countries its Image-100 satellite data processing facility, its paleontological and petrographical laboratory, and its metallurgical laboratory. The Landsat receiving station to be established at Hyderabad, India, will provide imagery at nominal cost to neighbouring countries. Finally, the Royal Observatory in Hong Kong continues to produce forecasts of typhoon tracks by computer, which are made available to other countries.[23]

C. The Role of International Agencies

In addition to such regional forms of cooperation in science and technology, there is also interregional cooperation. However, most developing countries face enough difficulties in cooperating even regionally and they often lack sufficient capabilities or even a perceived need to cooperate on a global scale. For this reason, various United Nations agencies have attempted to aid the developing countries by sponsoring cooperative projects in science and technology among such countries. Aid has usually taken the forms of financial assistance, technical and feasibility studies undertaken by UN experts, or the establishment of regional training, research, and information centers and networks.

As an example, in 1974 United Nations Development Programme experts made a feasibility study of organizing engineering consultancy services in Latin America and the Caribbean. This was followed by UNDP financing to help form the Working Group on Consulting Services, which was to undertake studies for regional planning, industrial programming, and negotiation of technology, to name but a few.[24]

In Africa, Asia, and the Arab region, the UN Environmental Programme and the Food and Agriculture Organization (FAO) have been cooperating with cotton-producing countries to plan and initiate an integrated pest-control program. In a similar vein, UNDP, the Rockefeller Foundation, and the World Bank are helping to create Cotton Development International, an intergovernmental research and development organization, to integrate and intensify the development of cotton in participating countries.[25]

Another example of an international agency supported and promoted by UNDP is Jute International, which was designed to undertake research, promotion, and marketing to increase the production and consumption of jute. The countries specifically involved in this venture are India, Nepal, Bangladesh, and Thailand.[26] Similarly, the network for farm mechanization in the Asia-Pacific region mentioned above is also supported financially by UNDP, which contributed some $550,000 during 1977–1979.

Through the creation of networks, the United Nations has been instrumental in establishing regional and interregional links. UNESCO has provided support to several regional and interregional centers, including microbiological research centers in Bangkok, Nairobi, Porto Alegre, Cairo, and Guatemala City and the International Center of Insect Physiology and Ecology in Nairobi.[27] The World Health Organization has supported TCDC activities of the Mediterranean Action Plan, calling for the establishment of a network of collaborating institutes, the development of studies and guidelines by scientists of the Mediterranean area, intercountry meetings, training activities, seminars, and workshops. The Universal Postal Union has established training centers at a cost of $216,138 in the Ivory Coast, Argentina, Thailand, Kenya, and Malawi, staffed in part by experts from other developing countries.

The International Civil Aviation Organization and UNDP have funded a multinational pilot and aircraft mechanic training center at Addis Ababa, similarly staffed, and the International Telecommunications Union has given support to multinational training centers in telecommunications such as the Multicountry P&T Training Center in Malawi for trainees from Botswana, Lesotho, Malawi, and Swaziland. The Central Food Technological Research Institute (CFTRI) in Mysore, India, has been supported by FAO and UNDP in the establishment and operation of an international training center in food and agricultural production, conservation, and processing technology.[28]

Through their studies and financial support, UN agencies have helped both to identify areas of interest among developing countries and to initiate projects. UN agencies do not, however, have the resources to support financially all such cooperative projects around the globe. Rather their primary role has been to locate areas of common interest among developing countries and to make technical feasibility studies so that these areas of common interest may be transformed into actual projects by the countries themselves. In the future, UN agencies, particularly UNDP, will significantly increase the scope and range of their activities on behalf of Third World cooperation as a result of the Conference on Technical Co-operation among Developing Countries and subsequent conferences, meetings and symposia. Just how this is occurring is discussed in the section on UN conferences below.

7. Impediments to Greater Third World Cooperation

Despite the number of projects begun, no one should be under any illusions that the path to greater Third World cooperation in science and technology will be easy. The obstacles in the way of such cooperation are formidable.

To start with, the extreme maldistribution of global research effort constitutes a major obstacle. With most of the world a scientific desert and some 95 percent of global expenditure on research and development concentrated in a handful of industrialized countries, it should be evident that developing countries have a very meager base on which to build and will of necessity be dependent in varying ways and to varying degrees on the scientific and technological metropolises of the world.

The problem is compounded by the fact that the richer nations, as the centerpiece of their efforts to stimulate more rapid economic growth in Third World countries, have spent the last thirty years in increasing North-to-South science and technology flows, for the most part in a dependence-inducing or -perpetuating manner. The sheer momentum thus created will not be easily diverted by more South-to-South flows or largely internal generation of technology, as has become quite clear from the 1979 United Nations Conference on Science and Technology for Development (UNCSTD), at which the industrialized countries rejected proposals for a truly meaningful redistribution of research and development on a global scale. There now exist in industrialized countries development agencies with substantial staffs and budgets and many individuals and organizational units within academic and research institutions with a vested interest in continuing these North-South flows because budgets must be spent and salaries earned.

An even more powerful obstacle exists in rich-country consumption patterns and in the insatiable appetites for certain kinds of consumer goods, including – as the general level of affluence rises in these societies – some of a socially quite trivial nature. Poor countries in search of foreign exchange (often for the wrong reason – namely, to import consumer goods for their largely urbanized upper-income groups) go where the markets are, acquiring whatever kinds of technical skills are necessary to serve those markets. But these skills are rarely of the right type or acquired in such a manner as to provide a basis for further self-generating economic growth within poor countries.

This leads to a very complicated and inhibiting political situation. Governments are made up of individuals who benefit from, and represent those who benefit from, the political economy in being, as Frances Stewart has observed. They may not wish or be able to challenge the existing political economy. An alternative technology, reflecting an alternative development strategy with different priorities at the national level, involves an alternative political economy – a very different distribution of the benefits of the economic system. Governments rooted in one system may not be powerful enough to choose an alternative system.[29]

The systematic character of technological development, with productive activity based on one kind of technology frequently requiring complementary activities based on similar technologies, is yet another difficult obstacle to overcome. Alternative technologies designed to meet the needs of those segments of the population now largely excluded from the benefits of technological change in Third World countries would also raise serious problems of the efficiency of the existing system of productive activity, directed as it is toward serving the needs of the small modern sector of the economy. The pervasive influence of these linkages of the modern sector of poor countries' economies with industrialized countries spills over into the economic infrastructure and social services that are required to operate a productive system based on advanced-country technology. The result is that most of the resources available to build up and maintain such infrastructure and services are absorbed by the modern sector, serving 10 to 20 percent of the population, leaving the traditional sector, with most of the people, largely neglected.[30]

As if these obstacles were not enough, there is the fundamental fact of political and economic life in the modern world. To paraphrase George Orwell's dictum in *Animal Farm*, all sovereign nations are equal, but some are more equal than others. One of the major reasons why efforts at economic integration with the Third World in Central America, the Andean region, West Africa, East Africa, and Southeast Asia have shown such limited progress, according to Arthur Lewis, is that some countries are more advanced than others and benefit more from integration at the expense of the others. The result is obvious: The agreements on which such efforts at regional integration are based are unstable. And yet regional economic integration is a critical element in achieving greater scientific and technological cooperation within the Third World.

This diversity is not confined to levels of economic development but exists also in cultural, ideological, and political fields as well. Wide variations in size are yet another element of diversity. Consider the Economic Community of West African States (ECOWAS), formed in 1975 when fifteen heads of state from countries as different as Ghana, Guinea-Bissau, Liberia, Mali, Nigeria, and Senegal met to sign the ECOWAS treaty.[31] The remarkable thing is that so much regional cooperation has occurred, not that there has been so little.

Finally, there is the inhibiting consequence of decades, if not centuries, of skewed development of the international economic infrastructure along North-South lines to the almost total exclusion of South-South links. It is still easier in many parts of the Third World to travel between developing countries via Europe or North America than it is to do so directly. The same condition applies to moving money, ideas, and goods within the Third World, making the task of strengthening Third World cooperation in science and technology and many other fields all the more difficult.

8. Countervailing Opportunities and Forces for Third World Cooperation

On the other side of the ledger are some very important considerations that should move Third World countries in the direction of greater cooperation among themselves scientifically and technologically as well as economically and politically. The most important countervailing force is a political one, which has already led to the emergence of the Group of 77 as a major bargaining instrument within the international community on North-South issues. Northern political commentators have repeatedly expressed disbelief that such a diverse group of nations with such diverse interests can find enough common ground to maintain a united front in negotiations with the industrialized countries. But developing countries, being much weaker actors in the international system, have long since learned that divided they will surely fail. Until their own political and economic power increases, solidarity is indeed their best hope in negotiations with the more powerful North.

Virtually all countries, rich and poor, seek to minimize their dependence on external forces and factors. Thus even such a powerful member of the international community as the United States is actively committed, at least at the rhetorical political level, to diminishing its energy dependence on the OPEC countries. Greater Third World cooperation in science and technology thus becomes an instrument for diminishing dependence, collectively and individually, on the North and acquiring more room for maneuver on the part of weaker elements in the international system.

Some developing countries, because of their very limited (in some cases virtually nonexistent) scientific base and problem-solving technological capacity, are going to be dependent on external help in coping with their own problems for some time, but they are more likely to receive socially and economically relevant, lower-cost solutions to these problems if they turn to other developing countries than if they continue to rely almost exclusively on advanced-country technology. It is important to emphasize that we are talking about matters of degree and not of kind. Obviously for solutions to certain kinds of problems, the most advanced techniques in the world will also be the most appropriate. The difficulty at present is that Third World countries rely, when they seek external solutions to their problems, almost exclusively on such advanced techniques.

But some Third World countries are beginning to accumulate a significant body of experience with alternative techniques that may be more appropriate to the needs of other Third World countries.

Most Third World political leaders, notwithstanding the contrary assertions of Northern political commentators, do want a better life for *all* their people and are genuinely anxious to eliminate suffering and deprivation within their countries. In all too many cases they lack the resources and the problem-solving capabilities to do so. They are also, of course, constrained by the particular kind of political economy, discussed above, that brought them to power. We cannot be certain that there is a causal link between greater technological self-reliance or autonomy and greater social equity, but there does appear to be a significant correlation. As we have suggested, few countries have been able to meet the minimum material needs of most, if not all, of their population on a sustained basis without substantially increased technological autonomy. Here also, then, is another countervailing factor in moving toward greater Third World cooperation in science and technology.

There are, furthermore, important global economic trends identified by W. Arthur Lewis and briefly discussed above that are likely to lead to greater South-South trading and technology links as access to Northern markets for Southern manufactured goods becomes more difficult.

Finally, there is yet another possibility, however remote it may seem from the vantage point of short-term political and economic realism. Enlightened leadership and potentially important political and economic forces in the North may perceive that their long-term interests will be better served by working to create an international system of more genuinely equal and autonomous actors better able to meet the needs of their own people and therefore less dependent upon richer nation-states in the system. To the degree that such leadership and such forces exist and can develop more meaningful linkages with similar leadership and social movements in the South, a new kind of alliance may be formed through which South-to-South cooperation in science and technology will be actively supported and encouraged by such elements in the North.

This is, to be sure, a visionary perception. But the rampant spread of social pathologies generated by indiscriminate technological change in the North, coupled with an accelerating equity crisis in the South and increasing threats to our frail biosphere for all people everywhere, may give increased impetus to such leadership and social forces. Thus, within the North, smaller and more progressive industrialized countries, now themselves in a highly dependent and vulnerable relationship to the larger industrialized countries, may find their national interests better served by strengthening linkages with some Third World countries. And within industrialized countries, trade unions with a strong commitment to social justice, consumer and environmental groups, appropriate technology organizations, and similar citizen action groups committed to pursuing economic and political alternatives for industrialized societies may achieve greater political clout than they now have.

Visionary as may be this view of possible future constellations of social and political forces in the world, such a constellation is much more likely to move rapidly toward those overriding goals of the United Nations of peace and social justice than is perpetuation of the present pattern of international economic and political relationships, which are deeply embedded in and reinforced by the global maldistribution of problem-solving capacity and prevailing international flows of science and technology.[32]

9. UN Conferences and Third World Cooperation

The Conference on Technical Co-operation Among Developing Countries (TCDC) held in Buenos Aires from 30 August to 12 September 1978 appears to be simultaneously a distinctive historical event – the coalescence of the efforts of a group of weaker and poorer countries to attack their dependence on more powerful actors in the international system through technical cooperation among themselves – and but one of many similar events in the larger historical framework. Combining both its distinctiveness and its commonality with other events in a single metaphor, the Conference may be seen as a swelling in a historical stream that reaches back to the end of World War II and stretches far into the future.

That aspect of this stream with which we are most concerned is the emergence and development of the so-called new nations since the Second World War. This development is twofold: both a material transformation of these countries characterized by economic and technological development and a "conative" transformation – that is, the striving to assert their own purposes, strategies, policies, and goals for development. Thus, the TCDC Conference must be viewed as an integral part of the effort of poor countries to develop from an economic and technological standpoint over the past several decades and of their continuing struggle to define and achieve their own plans and goals in this development.

In this latter sense, TCDC is part of the same genre of global events as the various UN Conferences on Trade and Development (UNC-TAD), the quinquennial conferences of the UN Industrial Development Organization, special sessions of the UN General Assembly on development, and UNCSTD. Thus, it is by means of forensic diplomacy, i.e., diplomacy undertaken by open debate in an appropriate forum or conference, that developing countries have sought to affect events and their own destinies.

One should not lose sight of the distinctive character of the TCDC Conference, however. In their quest for greater identity and self-control in the political, cultural, and economic spheres over the years, developing countries have come to the realization that this quest must also extend to the sphere of science and technology. One result of this quest has been the formulation of the notion of "collective self-reliance." Collective self-reliance is not merely an ideological imperative, but also an expression of a fundamental conative transformation of developing countries – a transformation of their perceptions of global relationships and a transformation of their purpose and resolve to affect these relationships accordingly. This was stated as follows in the Kuwait Declaration on TCDC of 1977:

> TCDC is a historical imperative brought about by the need for a new international order. It is a conscious, systematic and politically motivated process developed to create a framework of multiple links between developing countries. Many steps have already been taken in this direction. TCDC may be facilitated, or hindered, but it cannot be stopped.[33]

The significance of such a conference lies in both its educational and its instrumental character. Its educational significance results from the Conference's highlighting of key issues and problems involved in working toward Third World cooperation and from the propagation of the notion of collective self-reliance, a notion that is the cornerstone of the final TCDC plan of action. The instrumental value of the Conference will be determinably its ability to generate action in promoting TCDC. This implies the building of a workable consensus during the negotiations at the Conference both among the developing countries themselves and between the developing and the developed countries. Therefore, the key to understanding the negotiations at the Conference lies in the attempt of the participants to achieve such a consensus on the institutional mechanisms that would make operational the recommendations of the Conference embodied in the Plan of Action.

The dynamics of this process can be seen in part by examining the transformation of the draft plan of action into its final form. The recommendations of the draft plan are divided into three sections: recommendations for action at the national level, the regional level, and the global level. As the term TCDC implies, action at the national level means action that can be taken by the developing countries themselves to initiate, promote, and strengthen TCDC activities. As such, there is little direct action that can be taken by industrially developed countries. Because there is no necessary conflict of interest between the North and South at this level, the final Plan of Action is almost identical in terms of formulation and phrasing with the draft plan of action. As there was virtually no change in this section, it may be deduced that the developing countries were able to decide among themselves with little serious conflict the actions they were to take at the national level.

Recommendations at the regional level were expanded and made more explicit in the final plan as compared with the draft. Interregional cooperation was also emphasized. Again, there was little North-South conflict of interest in regional matters, and the general tenor of the draft plan was retained and its basic structure was unchanged in the final form.

Only at the global level is there evidence of significant change in the recommendations from the draft to the final plans. As it is at the global level that the developed countries have a larger role to play in TCDC, especially in financing TCDC activities, it follows that changes in the Plan of Action at this level are indicative of fundamental differences between South and North. A textual examination of the final and draft plans of action reveals that the difference between the two versions lies in the final plan's specific description of the global institutional arrangements that would help to operationalize the TCDC recommendations and provide financing for TCDC programs. Empirical evidence moreover supports the observation that over these issues there was greater disagreement between industrially developed and developing countries.

According to Chakravarthi Raghavan, industrialized countries, such as the United States, West Germany, and the Soviet Union, insisted upon having the UNDP Governing Council act as the parliamentary body for TCDC.[34] Because of their weighted representation, these countries would be assured of a measure of control of TCDC activities through UNDP. Developing countries, on the other hand, insisted upon an intergovernmental body, open to all Member States with equal participation rights and reporting directly to the UN General Assembly, which would give them effective control. A special unit or part of the UNDP Secretariat would provide the servicing machinery for TCDC programs.

Not until virtually the last minute was a compromise reached. The intergovernmental body that would be charged with reviewing and carrying out the Plan of Action would be "entrusted by the General Assembly to a high level meeting of representatives of all States participating in the United Nations Development Programme."[35] When this body reported to the General Assembly through the UNDP Governing Council or the Economic and Social Council of the UN (ECOSOC), the Governing Council and ECOSOC would be able to comment on the report and make their own recommendations, but could not modify or bottle up the report. As such, the developing countries, if they utilize this body properly, can "virtually bring the development activities of the entire UN system directly under their purview and exercise an overview of the UN and its sprawling specialized agencies."[36]

Concerning the financing of TCDC programs, the Plan of Action asks that industrially developed countries increase their voluntary contributions and accelerate the process of untying their aid resources.[37] In addition to this possible source, a percentage of the national, regional, interregional, and global Indicative Planning Figures (IPFs) of UNDP will be used to finance programs. This could amount to some $200 million or more annually. Already, India has indicated that it would use $5.8 million from its IPF for 1980–1981 for TCDC activity.[38]

In retrospect, from this limited distance in time, the TCDC Conference seems to have been among the more significant recent major UN conferences. In terms of its educational role, major issues and problems were brought to light, and the notion of collective self-reliance was given greater visibility and put into a historical context. From an operational perspective, a *potentially* effective institutional mechanism was created to implement the recommendations of the TCDC Plan of Action with the minimum of interference from the industrially developed countries and the international bureaucracy. The principal weakness of the Conference still lay in its lack of a specific requirement for financial goals to be met by Member States of the United Nations. Even UNCSTD, which had such a goal, faces financial difficulties, as will be seen. As always in assessing the actual performance of international institutions, the proof of the pudding lies in the eating.

Less than a year after the TCDC Conference was held in Buenos Aires, another major conference was held in Vienna, from 20–31 August 1979: the United Nations Conference on Science and Technology for Development (UNCSTD). From the outset, conflict was expected on a larger scale than at the TCDC Conference between the industrialized and developing countries, particularly over such issues as a code of conduct for technology transfer, patents, institutional mechanisms within the UN system, and international funding.

Developing countries were better organized at Vienna than at Buenos Aires, however, having agreed earlier on a position at the meeting of the Group of 77 in Bucharest. They demanded, *inter alia*, that a funding mechanism be created that would raise $2 billion by 1985 and $4 billion by 1990 by means of automatic, compulsory payments from industrialized countries, to be determined by a fixed percentage of the surplus from their balance of trade in manufactured goods with developing countries. An intergovernmental committee open to all countries would be established to govern the financing system and oversee all science and technology activities in the UN system. This committee would report directly to the General Assembly, thus insuring Third World control.[39] Most industrialized countries opposed these demands. Not until virtually the last moment, just as at Buenos Aires, was the impasse broken. But it was broken for the most part in favour of the industrialized-country position.

The Conference finally adopted the proposal that the Committee on Science and Technology for Development within ECOSOC be abolished and replaced by the new Intergovernmental Committee on Science and Technology for Development. This Committee was to be open to all states and would "submit its reports and recommendations to the General Assembly through the Economic and Social Council, which may transmit to the Assembly such comments on the report as it may deem necessary" but which may not alter the content of the report, thus retaining the essence of the original proposal of the Group of 77.[40]

A funding mechanism was also agreed upon, far removed from the Group of 77 proposals outlined above. Starting in 1982, the Intergovernmental Committee will direct the financial system for science and technology for development, its exact nature to be determined by then. Until that time, an Interim fund has been created, administered by UNDP and based on voluntary contributions, intended to amount to no less than $250 million for the period 1980 and 1981.[41] The results of the initial pledging conference held in the spring of 1980 suggest again the wide gap between international conference rhetoric and subsequent performance by national governments. Even though some more pledges are expected, the initial figure for the Fund is some $35 million, a far cry from the $250 million solemnly promised the preceding August.

The effectiveness of this fund, moreover, remains to be proven. Will governments that contribute to this fund then subtract funds accordingly from what they would have otherwise given to UNDP, UNESCO, and other international agencies for science and technology programs? If this happened, then there would be no net increase in funding projects for such programs. Alternatively, will UNDP, for example, no longer support projects dealing with science and technology in its regular programs, referring these projects instead to the Interim Fund? If this happened, then actual funding for such projects in the UN system would decrease.

With the heated debate on funding mechanisms and the like, the issue of cooperation among developing countries in science and technology for development was given less attention, but not forgotten. The final resolution for the Vienna Conference included proposals that developing countries should:

   a. Promote mutual consultation and systematic exchange of information concerning their experience in science policy and planning, building scientific and technological infrastructure, and the acquisiton, development and application of scientific and technological knowledge;

   b. Strengthen the existing and establish, develop and promote new consulting firms and services relevant to the area of science and technology;

   c. Make arrangements to facilitate the dissemination and exchange of science and technology knowledge and experience originating in the developing countries so that the comparative advantages and specializations of various countries or sectors can be fully utilized;

   d. Arrange for the training and exchange of science and technology personnel;

   e. Whenever possible, establish associations of reearch councils and joint research and development centers in areas of common interest, and provide for the exchange of recently developed science and technology knowledge;

f. Promote science and technology projects among developing countries with similarities in natural and social factor endowments;

g. Compile inventories of their science and technology resources and capacities for collective self-reliance in science and technology for development, and encourage their exchange.[42]

UNCSTD and the Conference on TCDC, however, were not the only international fora to discuss the issue of cooperation among developing countries in science and technology for development. A less well-known conference was the World Conference on Agrarian Reform and Rural Development, sponsored by FAO and held in Rome, 12–20 July 1979. It adopted a Declaration of Principles that urged developing countries to strengthen their technical cooperation in rural development and to foster policies of collective self-reliance. It called upon developing countries to exchange their experience and expertise in agrarian reform and rural development, to exchange technologies among countries with similar natural conditions and social systems, and to promote intercountry rural projects, such as irrigation and watershed management.[43]

When viewed primarily from a North-South perspective, the substantive results of conference diplomacy may seem disappointing. Yet they do appear to provide a stimulus to more South-South exchanges and information sharing on technical and other subjects. Preceding and following on the heels of every major international conference is a plethora of smaller conferences, meetings, symposia, and workshops in the Third World to discuss in greater detail the issues of the major conferences as they relate to local or regional conditions. The flow of scientific and technical personnel into and among developing countries is thus facilitated as one consequence of organizing major world conferences on development problems.

10. Ideas and Proposals for Action

In this section ideas and proposals for action will be examined in order to suggest further possibilities for cooperation that have not yet been implemented on any large scale. In doing so, we distinguish between ideas for action and specific proposals. Specific proposals are usually self-explanatory and refer to a concrete action to be taken by specific persons or countries. An idea for action, on the other hand, is more general and typically refers to general guidelines or general conditions that make a specific proposal possible.

In discussing Third World cooperation in science and technology for development, ideas for action and specific proposals take two forms: cooperation by countries in implementing various types of projects and the creation of institutional mechanisms that facilitate such projects specifically and make greater cooperation possible in general. These institutional mechanisms encompass such activities as information systems, educational and training programs, and so forth.

There are many instances of specific proposals for Third World cooperation in different fields. Interest in this topic is not a new phenomenon. For example, almost fifteen years ago, at the Symposium on Collaboration Between the Countries of Africa and Asia for the Promotion and Utilization of Science and Technology held in New Delhi in 1966, a number of specific proposals urged African-Asian cooperation. S. P. Raychaudhuri mentioned the possibility of international cooperation in land reclamation and development in wastelands, in techniques for controlling wind erosion, and in the optimum utilization of soil moisture under desert and semidesert conditions.[44] Other suggestions included collaboration in plant protection, mineral research, soil sciences, and petroleum and gas geology. As was indicated above, a number of such specific project proposals have since been implemented.

Of greater theoretical interest are ideas for action that help to create the conditions making possible or facilitating cooperation on specific projects. Francisco Sagasti, in his paper for the Pugwash Symposium in Tanzania in 1975, suggested that such cooperation is possible in the following cases:

- Activities which require a minimum critical mass to be performed. This includes research and development for which it is necessary to count with professionals, equipment, and financing at a level below which the activities are not viable. In these fields it is impossible to intervene individually and cooperation efforts are indispensable.

- Scientific and technological activities in which there are economies of scale (information systems, training programmes, etc.). In this case international cooperation is not absolutely necessary, but involves many benefits which make it highly desirable.

- Activities which must involve an international dimension to make sense. This includes comparative and joint actions that are meaningless when considered in only one country. An example would be the establishment of comparative information systems on terms and conditions for technology transfer, which would increase the bargaining power of the countries buying technology. This could be extended to agreeing on common strategies for negotiations with technology suppliers and to the adoption of common positions before multinational corporations, multilateral financial institutions, and other similar organizations.

• Problems common to more than one country, linked to geographical zones that extend beyond national frontiers. This includes research into ecological conditions, the exploitation of natural resources, use of water systems, and so on. In this case the existence of a common problem provides the countries with the possibility of joining forces in the performance of scientific and technological activities.

• Large undertakings in which it is necessary to share risks among several countries because of the magnitude of resources required. This has been the case of investments in nuclear energy, computers, satellite telecommunications, etc., in which few individual countries—even if they are capable of financing the programme on their own—are willing to take the risk alone.[45]

In addition to the various forms of specific projects that may be undertaken by developing countries in concert, there are also several forms of institutional mechanisms that may either facilitate such projects in particular or may facilitate cooperation among developing countries in general. These mechanisms include the following: (1) sharing of information, knowledge, and experiences; (2) creating professional organizations; (3) cooperation in education, training, and research and development; (4) linking policies and pooling resources; and (5) creating financial mechanisms.

A. Information Exchange

Information exchange and the opening of channels of communication between developing countries is seen as an important step toward enhancing Third World cooperation in science and technology for development. The Group of 77 adopted this as a working principle in the Kuwait Declaration of June 1977: "TCDC must promote and sustain networks of information and of cooperation and for coordination between specialized institutions and organizations of developing countries."[46]

Information consists of both knowledge and experiences. The Pugwash workshops at Badkal Lake, India, in January 1978 and at Rabat, Morocco, in April 1978 drafted the following proposals for such forms of information exchange:

a. Promote the systematic exchange of information concerning their experience in science policy and planning, building S and T infrastructure, and the acquisition, development and application of S and T knowledge. In this regard regional and sub-regional information centers could be set-up.

b. Establish machinery to facilitate the dissemination and exchange of S and T knowledge and experience originating in the LDCs so that the comparative advantages and specializations of various countries or sectors can be fully utilized.[47]

At the Arusha Symposium of 30 January to 4 February 1978, scientists and development experts stressed the importance of documentation centers and freedom of information:

a. African Member States of the United Nations Economic Commission for Africa (ECA) should expedite the arrangements for setting up a computerized documentation center for the African region with branches in as many states as may be possible.

b. Several copies of every publication . . . in or about the African region must be made available to the proposed Regional Documentation Center by the author . . . or publisher(s) free of charge, for distribution to national libraries and for microfilming and storage in the data bank of the proposed Regional Documentation Center.

c. African Member States of the UN/OAU, in collaboration with the ECA and other UN specialized agencies, should press for an International Freedom of Information Act to enable the proposed Regional Documentation Center to have direct access to such sources of information as the U.S. Library of Congress, etc.[48]

Such proposals were also emphasized in many of the national papers for UNCSTD. For example, Ethiopia stressed the need for a regional information and documentation center.[49] The Kenyan National Paper also stressed the need for greater communication among developing nations. For example, "each developing country should provide information on its market opportunities for other developing countries" and "should take a comprehensive inventory of its S and T capabilities to facilitate identification of common areas of co-operation."⁵⁰

The UNCSTD Programme of Action adopted in Vienna called for the creation of a global scientific and technological information network making widely available "data from the developed and the developing countries on available technologies, conditions of licensing, identification of suitable experts, engineering and consultancy services and the like."⁵¹

B. Professional Organizations

The creation of professional organizations or the enhanced use of existing organizations may facilitate information exchanges, build up infrastructures, and create an atmosphere to further cooperative action. The Arusha Symposium gave several proposals along this line:

a. Regional professional groups in all disciplines should be formed. They should be encouraged by national scientific councils and other government agencies to publish their works freely in journals started by these groups in their respective disciplines.

b. African governments should encourage and support professional associations at the national level and the relevant UN agencies and other international organizations should give such support at the regional level.

c. African National Scientific Research Councils should form an association which could bring them together regularly for symposia, etc., and for the promotion of regional co-operative research programmes.⁵²

Where such institutions do exist, participating countries have sometimes found them significant enough to cite them in their UNCSTD national papers. For example, the Indonesian National Paper for UNCSTD stated that regional bodies such as the Association of Asian Institutions of Higher Learning have been helpful in their development.⁵³

C. Education, Training, and R&D

The first two mechanisms mentioned above seek to establish cooperation among developing countries by creating, in most cases, new institutions that cut across national boundaries. But existing institutions in developing countries may also be linked, thereby enhancing cooperation. Links may be created between educational, training, and research institutions. For example, the Arusha Symposium suggested that:

> African universities and research institutions should strengthen their links among themselves as well as with overseas institutions, in order to exchange information, staff and students and to launch cooperative research programmes and dialogues in such areas as solar energy technologies, materials for low-cost housing for rural areas, health, nutrition, etc.[54]

The Pugwash workshops have also suggested that developing countries should:

a. Make appropriate institutional arrangements for the training and exchange of S and T personnel.

b. Establish associations of research councils and joint R and D centers in areas of common interest, and machinery for exchange of recently developed S and T knowledge.[55]

Similarly, the Pakistan National Paper for UNCSTD emphasized the creation of a network of cooperating institutions among developing countries in order to promote joint research.[56]

D. Linking Policies and Resources

Cooperation of this sort was mentioned by the Pugwash workshops in a general form for developing countries to:

a. Endeavor to co-ordinate their S and T policies.

b. Pool their S and T resources and capabilities for collective self-reliance in S and T for development.[57]

The Kenyan National Paper was more specific in its proposals. Developing countries in a region should establish lines of communication among themselves and then provide information on each of the scientific and technological capabilities and market opportunities of each country. With this information in hand, each country would specialize in the production of goods and services in which it had a comparative advantage. Needless to say, the Kenyan paper pointed out that "a lot of political goodwill and commitment is required for this co-operation."[58]

E. Financial Mechanisms

Financial mechanisms are the topic of Chapter 4 by Francisco R. Sagasti. The mobilization of financial resources is so crucial to the possibility that anything of significance might actually happen that the topic cannot be ignored. But we will make only brief mention here because of the detailed discussion in Sagasti's chapter.

National governments may undertake a number of measures to finance their science and technology programs. These include:

a. The so-called ITINTEC system whereby all industrial enterprises in a country or region, regardless of their ownership, must set aside 2 percent of their net income before taxes for technological research in the country or region in question. Sagasti estimated that this mechanism alone could generate US$552 million in the Caribbean and Latin America, US$96.4 million in the Middle East, US$61.3 million in Africa, and US$243.3 million in East and South-East Asia.

b. The payment into a fund to strengthen local research and development of an amount equal to payments for royalties and other fees for foreign technology. Sagasti estimated that more than US$500 million could be generated annually in Latin America through this mechanism.

c. The setting aside of one-fourth of 1 percent of loans for industrialized development given by development financing institutions to industrial enterprises. Argentina currently uses this mechanism.

d. The requirement that subsidies of transnational corporations spend in S&T activities in the host countries the same percentage of sales as they spend in their home country. For U.S. firms in Latin America, such a mechanism could generate some US$500 million.

e. The requirement that some percentage of direct foreign investment flows be set aside for local R&D. For a 1 percent requirement, this would generate some US$95 million in Latin America for a six-year period.[59]

In addition to these national sources, mechanisms tapping international sources could be created, such as the following:

a. Specifying 5 percent of bilateral aid for S&T development purposes. Special-purpose institutions, such as the Canadian International Development Research Center (IDRC) and the Swedish Agency for Research Co-operation with Developing Countries (SAREC), could be used to channel these portions of bilateral development assistance earmarked for S&T development.

b. Using private agencies and instutitions for developing the autonomous S&T capabilities of developing countries. These include the Ford Foundation, Rockefeller Foundation, Friedrich Ebert Foundation, International Foundation for Science, etc.

c. Using multilateral financial agencies, whereby 10 percent of all their loans could be specifically allocated to the development of endogenous S&T capabilities of developing countries. This could generate some US$450 million per annum. In addition to UN agencies, the World Bank and the various African, Asian, and European development and investment banks might be used for this purpose.[60]

F. Multifunctional Institutional Mechanisms

A number of the foregoing ideas and proposals are encompassed in the set of proposed initiatives developed as part of the preparatory process for UNIDO III. Sanjaya Lall has called this set of propositions an "International Technology Brokerage System."[61] Peter O'Brien and his coauthors, in a companion paper, identified its principal components as an Agency for Technology Financing, Insurance and Trading, a Third World Corporation for Advances in Technology, and an International Technology Brokerage Organization, which together would facilitate South-South flows of technology through the collection and dissemination of technical information, matchmaking between buyers and sellers of technology, financing and technology purchases, and provision of support activities, such as engineering consultancy and design services.[62]

Along similar lines, the Group of 77 adopted at its meeting in Bucharest just before UNCSTD the financial proposal that a funding mechanism be created that would raise $2 billion by 1985 and $4 billion by 1990 by means of automatic, compulsory payments from industrialized countries of a fixed percentage of their balance of trade in manufactured goods with developing countries.[63] As indicated above, however, this proposal was opposed by the industrialized countries and was not, therefore, adopted in the Vienna Programme of Action.

Of course, there are many more examples of ideas for action and proposals for both specific projects and for institutional mechanisms for implementing these projects.[64] This account merely attempts to suggest the wealth of possibilities not yet widely implemented and, as in the case of the TCDC Conference, recommendations perhaps soon to be implemented.

11. Outlook For A New International Science and Technology Order

It is sometimes suggested that the various recent and forthcoming international events dealing with the changing role of science and technology in the international system should define the scientific and technological component of the Third Development Decade, introducing a New Scientific and Technological Order to parallel and help accelerate achievement of the New International Economic Order. Something certainly needs to be done to help the NIEO along, for there has been depressingly little progress since it was promulgated at the Sixth Special Session of the General Assembly in 1974. But it is hard to see how international events like UNCSTD or TCDC, dealing with a dependent variable in the process of economic and political change, can do much in the face of manifest lack of interest on the part of the major industrialized countries in rewriting the international rules of the game along more equitable lines.

The near-failure of the Conference on International Economic Cooperation in Paris, the persisting difficulties in negotiating a common fund for international commodity trade stabilization of meaningful scope and character, and the overt lack of agreement on concrete steps to achieve global distribution of industrial production at UNIDO III in New Delhi in early 1980 illustrate how hard it is to achieve meaningful progress where substantial modification in industrialized-country policies and attitudes is required. There are simply too many powerful vested interests within these countries to make more than superficial changes possible.

Some industrialized countries have by now, furthermore, acquired a knee-jerk reaction to Third World demands for the NIEO, refusing to listen seriously when demands are voiced that to the former group of countries imply redistribution of the world's existing wealth. Resistance immediately begins to harden when developing countries seek *significant* access to desired technology and markets on equitable terms unless they are bargaining from a position of political and economic strength. Preferential terms for developing countries such as generalized system of preferences (GSP) schemes on tariffs or technology transfer have thus far been only crumbs off the table of industrialized-country riches, not meaningful concessions.

This is not an argument against continuing the effort to make established international political and legal rules less discriminatory in areas such as investment, technology transfer, and patents. On the contrary, these efforts have a salutary educational impact in industrialized countries by repeatedly underscoring how discriminatory the existing rules are. It is rather an argument for realistic recognition of what can be accomplished. OPEC has demonstrated that only where real bargaining power is present can substantial concessions be achieved. And even the OPEC advantage may prove to be more illusory than real because of the rapidly accelerating technological dependence of most of the oil-exporting countries on the major oil-importing countries.

Most Third World countries are confronted with a Hobson's choice in the coming decades. Third World countries can opt for increasing integration in the world productive system as highly dependent actors subject to the superior economic and technological capacity of the more powerful industrialized countries, which will, when push comes to shove, act to protect their own interests. This may yield higher rates of growth in the short term, but these are likely, if past experience is any guide, to be accompanied by even greater maldistribution of income and higher rates of unemployment.

On the other hand, Third World countries can strike out in a more autonomous way, attempting to build their problem-solving capacities by tackling their problems in the first instance by themselves and relying only secondarily on external help, when they can get it with a minimum of entangling conditions. This will almost certainly mean slower growth rates, although quite possibly more equitable patterns of growth, and the gap between the rich and the poor countries will doubtless increase in the remaining decades of this century. But if developing countries opt for this strategy of development, collective self-reliance based on Third World cooperation can play a potentially important even if in the beginning modest role along the lines suggested in this chapter.

In many ways collective self-reliance and the quest for greater na-

tional autonomy are "least worst" strategies. But as Winston Churchill remarked about democracy as a form of government, it may be
the worst there is, except for all others.

12. A Question of Balance

Clearly Third World cooperation in science and technology, and
more broadly technical cooperation among developing countries,
are ideas whose effectiveness has yet to be tested on any scale. To
meaningful alternatives to conventional North-South science and
technology flows, their momentum will accelerate. And there is
already enough evidence of actual experience with such cooperation
to show that it can work, given the right circumstances and a commitment to making it go by the parties concerned.

On the other hand, there are substantial obstacles, as we have
noted. Perhaps the potentially most important of these is the prospect of Northern interference if the idea really begins to gather
momentum. Even at this very early stage in the evolution of collective self-reliance within the Third World, the Northern-controlled
mass media have done a remarkably effective job of pretending that
South-South cooperation either does not exist or never ac-
complishes anything of any consequence.

It is probably safe to say that Third World cooperation in science
and technology will not accomplish as much as its most ardent advocates claim for it. But it may very well accomplish much more
than its detractors, mostly in the North, predict. The political
economy in the world in the year 2000 will surely be quite different
than it is in 1981, and one dimension of that difference will be far
more meaningful South-South institutional arrangements and relationships in science, technology, and many other fields of human
endeavour.

This assessment of the recent TCDC Conference by an independent Indian journalist and observer, Chakravarthi Raghavan, sums
up the situation not only for TCDC as a whole but for collective
self-reliance in science and technology as well:

The Plan of Action for TCDC was at best a signpost. Whether
TCDC would really become a new concept for collective self-reliance
of the South, or merely turns out to be "travel costs for Third World
countries," would really depend on how the idea is further developed
and promoted. TCDC can become a movement towards the goal only

when the Third World realizes that NIEO will not be given to them or fall on them like manna from heaven, but has to be created by their own efforts.[65]

Notes

This chapter is based in part on a background paper prepared by David W. Chu for the September 1978 NGO Forum on Science and Technology for Development and on papers by Ward Morehouse prepared for the Jamaica Symposium on Mobilizing Technology for Development (January 1979) and the Lund Seminar on Science and Technology in the Changing International System (May–June 1978).

1. Zbigniew Brzezinski, "The World According to Brzezinski" (Interview with James Reston), *New York Times Magazine*, December 31, 1978, p. 10.

2. For further discussion of the unholy alliance of power, privilege, and technology in North and South, see Chapter 2, as well as Frances Stewart, *Technology and Underdevelopment* (London: Macmillan, 1977).

3. On the lack of credibility of global wars on poverty, see "Delinking North and South: Unshackled or Unhinged?" in C. F. Diaz-Alejandro et al., *Rich and Poor Nations in the World Economy* (New York: McGraw-Hill for the Council on Foreign Relations 1980s Project, 1978).

4. Some critics of the strategy of increasing integration of the modern sector of Third World economies into the global production system dominated by industrialized-country TNCs carry the argument further by pointing to an accelerating "de-industrialization" of the Third World, which is already occurring at an alarming rate and bids fair to get worse. Extraction of mineral resources and the increasing reorientation of Third World agriculture toward production of cash crops and nonfood products are in many cases driving the poor from the land. Environmental degradation, which is destroying the productive capacity of the land, is making the problem still worse. See, for example, Claude Alvares, "Development against People," *Development Forum*, July 1978.

5. For a discussion of the elements of a policy of technological disengagement, see Chapter 2.

6. W. Arthur Lewis, *The Evolution of the International Economic Order* (Princeton: Princeton University Press, 1977), p. 71.

7. Ibid. (emphasis added).

8. Ibid., pp. 74–75.

9. Sanjaya Lall, "Developing Countries as Exporters of Industrial Technology," p. 4 of unpublished paper (published in H. Giersch, ed., *International Economic Development and Resource Transfer* [Tübingen: J.C.B. Mohr, 1979]).

10. Ibid., pp. 5–7.

11. Ibid., p. 7.

12. Ibid., p. 9.

13. Ibid., pp. 10–12.

14. Ibid., p. 13.

15. United Nations Development Programme, *Technical Co-operation Among Developing Countries*, Case Study No. 15 (New York: UNDP, 1978).

16. UNDP, *Report on the Progress Made in Implementing the Tasks Entrusted to the UN Development System by the Buenos Aires Plan of Action for Promoting and Implementing TCDC*, First Draft by the Special Unit for TCDC, p. 36.

17. Ibid., p. 26.

18. Ibid., p. 32.

19. United Nations Conference on Technical Cooperation Among Developing Countries, *TCDC Case Study No. 11* (New York: Division of Information, United Nations Development Programme, n.d.), pp. 13–14.

20. Ibid., p. 24.

21. UN Conference on TCDC, *TCDC Case Study No. 14* (New York: Division of Information, UNDP, n.d.), pp. 2–3.

22. UN Conference on TCDC, *TCDC Case Study No. 7* (New York: Division of Information, UNDP, n.d.), pp. 11–12.

23. Ibid., p. 35.

24. UN Conference on TCDC, *TCDC Case Study No. 5* (New York: Division of Information, UNDP, n.d.), pp. 9–10.

25. UN Conference on TCDC, *TCDC Case Study No. 8* (New York: Division of Information, UNDP, n.d.), pp. 10–11.

26. UN Conference on TCDC, *TCDC Case Study No. 9* (New York: Division of Information, UNDP, n.d.), p. 16.

27. Ibid., p. 20.

28. Ibid., pp. 21–22.

29. Stewart, op. cit., especially pp. 274–279.

30. For a statement of the interlocking character of these linkages and their social consequences, see Amir Jamal, "Consequences of Accelerating Technology on the Least Developed Countries' Economies," paper prepared for Jamaica Symposium on Mobilizing Technology for Development, Ocho Rios, 7–11 January 1979 (published in revised form in Jairam Ramesh and Charles Weiss, Jr., eds., *Mobilizing Technology for World Development* [New York: Praeger Publishers, 1979]).

31. For an account of ECOWAS, see Barbara Harrell-Bond, "ECOWAS: The Economic Community of West African States," *American Universities Field Staff Reports*, 1979, No. 6 (Africa).

32. For a thoughtful and forward-looking statement of the possibility of new alliances following these lines, see Peter O'Brien, "The Eternal Triangle: Another Look At The Relationships Between Development, Industrialization, and Technology," paper prepared for International Workshop on Technological Dependence, Bonn, 2–5 November 1978,

especially pp. 31-36. The paper has been published in Dieter Ernst, ed., *The New International Division of Labour, Technology and Underdevelopment: The Consequences for the Third World* (Frankfurt: Campus Verlag, 1980).

33. *The Kuwait Declaration on Technical Co-operation Among Developing Countries* (Kuwait Ministry of Planning and UNDP, 5 June 1977), paragraph 2.

34. Chakravarthi Raghavan, "TCDC: Towards Collective Self-Reliance," *IFDA Dossier*, November 1978, p. 13.

35. *Report of the United Nations Conference on Technical Co-operation among Developing Countries*, A/CONF/79/131Rev.1, p. 19, paragraph 62.

36. Raghavan, op cit.

37. *Report of the Conference on TCDC*, p. 18, paragraph 60.

38. Ibid., p. 28.

39. *Retort* (NGO newspaper at UNCSTD), August 20, 1979, Issue No. 1, p. 1.

40. *Report of the United Nations Conference on Science and Technology for Development*, United Nations 1979, A/CONF/81/16, pp. 77-78.

41. Ibid., pp. 82-83.

42. Ibid., p. 69.

43. UNDP, *Report on the Progress in Implementing TCDC*, p. 15.

44. *Report on the Symposium on Collaboration Between the Countries of Africa and Asia for the Promotion and Utilization of Science and Technology*, New Delhi, 25 April-2 May 1966, p. A12.5.

45. Francisco R. Sagasti, "Notes on Technological Self-Reliance and Co-operation among Third World Countries," in W. K. Chagula, B. T. Feld, and A. Parthasarathi, eds., *Pugwash on Self-Reliance* (New Delhi: Ankur Publishing House, 1977), pp. 191-192.

46. *The Kuwait Declaration on Technical Co-operation Among Developing Countries* (Kuwait Ministry of Planning and UNDP, 5 June 1977), paragraph 6.

47. "Guidelines for International Scientific Co-operation for Development," *Pugwash Newsletter*, Vol. 15, No. 4, May 1978, p. 138.

48. Arusha Symposium, *African Goals and Aspirations* (Dar es Salaam: Tanzanian National Scientific Research Council, 1978), pp. 30-31.

49. Ethiopian National Paper, UNCSTD 1978, p. 34.

50. Kenyan National Paper, UNCSTD 1978, pp. 30-31.

51. *Report on UNCSTD*, p. 68.

52. Arusha Symposium, op. cit.

53. Indonesian National Paper, UNCSTD 1978, p. 44.

54. Arusha Symposium, op. cit.

55. "Guidelines for International Scientific Co-operation for Development."

56. Pakistan National Paper, UNCSTD 1978, p. 28.

57. "Guidelines for International Scientific Co-operation for Development."

58. Kenyan National Paper, UNCSTD 1978, pp. 30-31.

59. See Chapter 4.

60. Ibid.

61. Sanjaya Lall, "Third World Technology Transfer and Third World Transnational Companies," in *Industry 2000 – New Perspectives: Collected Background Papers* (Vienna: United Nations Industrial Organization, 19 December 1979) (UNIDO/IOD.326), Vol. 3, pp. 230ff.

62. Peter O'Brien, A. Hasnian, and E. Jimenez-Lechuga, "Direct Foreign Investment and Technology Exports Among Developing Countries: An Empirical Analysis of the Prospects for Third World Cooperation," in *Industry 2000*, pp. 185ff.

63. *Retort*, August 20, 1979.

64. A number of additional ideas for action and specific proposals were presented in the Report of the Task Force on Third World Co-operation at the September 1978 NGO Forum on Science and Technology for Development. The report was circulated by the NGO Committee on Science and Technology for Development, c/o Natural Resources Defense Council, 122 East 42nd Street, New York, N.Y. 10017.

65. Raghavan, *op. cit.*, p. 14. For a skeptical view of the consequences of TCDC, see Dieter Ernst, "Technical Co-operation Among Developing Countries (TCDC) – A Viable Instrument of Collective Self-Reliance?" paper presented to the DVPW Kongress, Augsburg, October 1979.

4
Financing the Development of Science and Technology in the Third World

Francisco R. Sagasti

1. Introduction

This chapter reviews some of the issues involved in financing the development of endogenous scientific and technological (S&T) capabilities in Third World countries and examines several mechanisms that have been established or proposed to this effect. After stating a few basic premises regarding the characteristics that financing mechanisms should have, the chapter discusses briefly the types of S&T activities to be financed, a variety of national financing schemes, and possible international funding mechanisms, suggesting in broad outline the arrangements that would be required to generate and channel additional funds. It concludes with a review and assessment of the Group of 77 proposals at the 1979 United Nations Conference on Science and Technology for Development (UNCSTD), as well as an examination of the agreements reached on financing matters.

It is now generally acknowledged that a massive increase in the amount of financial resources for endogenous scientific and technological development in the Third World is required. Estimates for 1973 indicate that world expenditures for research and development (R&D) were approximately US$100,176 million, of which the developing countries accounted for US$2,876 million, or less than 3 percent of the world total.[1] Furthermore, these disparities in financial allocations have been maintained for a long time, giving rise to cumulative differences of even greater magnitude. With such glaring imbalances in the distribution of the world scientific and technological effort it is clear that science and technology cannot contribute effectively to the self-reliant and autonomous development of Third World countries, and that a substantive redistribution of financial resources for science and technology is required.

A country's economy has to generate sufficient surplus to satisfy the consumption needs of the population and at the same time feed a viable process of accumulation to reproduce and expand the productive base. From an economic point of view, one of the key areas for investing the surplus is the development of S&T capabilities, which in turn would help further the process of capital accumulation through increased productivity and through the introduction of new products and processes. But the returns on such investments are seen only in the long run, are diffused throughout the whole productive structure, and are relatively more risky than alternative investment projects, such as the development of a physical infrastructure (roads, ports, power generation, etc.), agricultural development projects (irrigation schemes, drainage, etc.), and the establishment of industrial plants. For these reasons, investments in S&T are difficult to justify with traditional methods of project evaluation, and much work is required to develop appropriate project formulation, evaluation, and monitoring procedures for such a difficult area of investment. Therefore, given the prevailing economic concepts of project evaluation, often imposed by international financial agencies and organizations, it is perhaps too much to expect that at the early stages of development most Third World countries would allocate a significant share of their financial resources to the relatively uncertain task of endogenous S&T development. Furthermore, the urgency and impact of short-run problems usually preclude the possibility of focusing attention on long-term undertakings.

Nevertheless, Third World countries *must* make these allocations. There is no escape from the condition of underdevelopment unless endogenous S&T capabilities are built up.[2] This requires a change in the mentality and attitudes towards S&T, a reevaluation of S&T as factors of development, and the full incorporation of S&T considerations into the development planning process. Concurrently, a massive expansion of resources of all types, and of financial resources in particular, will be required for endogenous S&T development. However, considering the present economic difficulties of most Third World countries, it is highly doubtful that they will be able to generate on their own the necessary resources for such a massive expansion of funds for S&T, at least in the short term. Consequently, a transfer of resources from developed to developing countries must take place.[3]

It is also known that the expansion of the S&T infrastructure may lead to the development of "supply capabilities" with regard to scientific and technological knowledge, but that the crucial link with productive and service activities that characterizes endogenous S&T development does not emerge automatically out of the expansion of supply capability alone. It is also necessary to act, among other areas, on the pattern of demand for S&T knowledge. One of the key policy instruments in this regard is the financing of development projects, and attention should also be paid to the indirect S&T impact of development projects in general, proposing measures to ensure that financial practices lead to an increased demand for local S&T capabilities. However, this chapter will consider only those aspects of financing for endogenous S&T development that concern the performance of scientific and technological activities, understood in the widest possible sense.[4]

2. Characteristics of S&T Financing Mechanisms

There are several premises that have been generally accepted as guiding principles for the design and establishment of financial mechanisms for S&T, and it would be useful to review them before describing some of these mechanisms.

Financing the development of endogenous S&T capabilities is primarily a national responsibility and a task for the state in the Third World countries. To a large extent, this amounts to a recognition of the existing state of affairs, for government agencies in Third World countries account for the majority of S&T expenditures, given the combination of weakness and unwillingness of the private sector to participate actively in the development of endogenous S&T capabilities. Government intervention, particularly through financing mechanisms, is required also because of the relatively long-term nature of the task, the "externalities" involved, and the uncertainties associated with investments in S&T.[5]

When establishing financing mechanisms and deciding on the allocation of funds, it is necessary to define problem areas, programmes, and projects in which to invest, even though it is not necessary that detailed specific projects and programmes be always defined in advance. This is one of the most difficult problems to solve, and there have been cases in which resources were available but projects could not be formulated. Furthermore, when determining the level of financing for a certain set of activities, it is necessary to define at least two sets of parameters: the lower limits for financial allocations derived from the minimum critical mass necessary for the performance of S&T activities (in quantitative, qualitative, and interphase terms), and the upper limits, derived from the absorption capacity of the institutions involved in the performance of S&T activities (existing infrastructure and possibilities for expansion). This also underscores the need for a gradual and cumulative buildup of the human resources base and for a careful definition of priorities.

Financing should be provided for a broad range of S&T activities and not only for research and development, as has been traditionally done. In addition to research and development, S&T training programmes, adaptation of technology, search for technologies to import, disaggregation of the technological package, consulting and engineering design, risk capital to stimulate innovation, S&T information networks, quality control, and standardization registries of licensing agreements and patents, among others, are activities that need to be incorporated in a comprehensive scheme for financing endogenous S&T development. Furthermore, these activities require automatic, continuous, and predictable financing, preferably free from the vagaries of periodic budgetary negotiations and from the instability of voluntary contributions.

Closely linked to the broad range of S&T activities to be financed, there is a need for a multiplicity of institutional arrangements. Experience has shown that it is not possible to use a single scheme for identifying priorities and financing the whole range of activities and problem areas related to S&T development and that there is a need for a variety of sources of priorities for programmes and projects.

Cooperation among Third World countries at the subregional, regional and interregional levels is an essential component of endogenous S&T development. In addition to the need to achieve the minimum critical mass necessary in some fields of S&T, to benefit from economies of scale, and to confront the pressures of industrialized countries and transnational corporations, the fact that Third World countries share common perceptions of the problems of putting S&T at the service of development, a common historical legacy with regard to the lack of an endogenous S&T base, and also many problems for which there do not exist adequate S&T responses all contribute to make S&T cooperation among Third World countries an urgent imperative.

Finally, it is also necessary that the industrialized nations, in both East and West, support the development of endogenous S&T capabilities in the Third World. If the postulates of international equity and social justice that characterize the concept of a New International Economic Order are taken seriously, then an essential component of the efforts toward greater world equity and social justice is the development of endogenous S&T capabilities in the Third World. Given the resource limitations referred to earlier, and the additional fact that about one-half of the S&T effort of industrialized nations is devoted to the improvement of armaments and means of destruction, there is a strong justification for the industrialized countries to support the development of endogenous S&T capabilities in the Third World. Furthermore, the development of S&T in the Third World could also be in the interests of the industrialized countries, thus reinforcing the moral justifications for assistance with a component of "enlightened self-interest," as the Brandt Commission Report states clearly.[6]

3. Lines of Action for the Development of Financing Mechanisms

The expansion of financial resources for endogenous S&T development in the Third World requires simultaneous action along three different lines:

a. Identifying a coherent set of national, subregional, regional,

and interregional programmes and projects for endogenous
S&T development, to be financed through concerted national
and international action;

b. Restructuring the expansion of existing, and establishment of
new, financial mechanisms at the national level; and

c. Restructuring and expansion of existing, and design and
establishment of new, financial mechanisms at the international level, particularly to channel the support of industrialized countries for endogenous S&T development in
the Third World.

A. Identification of Programmes for Endogenous S&T Development

The identification and definition of priorities and projects for endogenous S&T development is not an easy task. Frequently the lack
of S&T capabilities precludes the clear identification of priorities
and the definition of projects, thus leading to a vicious circle: There
is no capacity to identify projects because there are no S&T
capabilities, and the latter cannot be developed because programmes and projects cannot be identified. Although external
technical assistance can help in many cases, it is not possible to rely
exclusively on it; otherwise the learning component involved in the
autonomous definition and execution of S&T projects would be
lost. It would also be necessary to undertake greater risks and
change the point of view of financing agencies (although it is *not* a
question of lowering standards), so as to allow for greater freedom
and flexibility on the part of the recipients, even if projects are not
formulated with the usual degree of detail to which development
bankers are accustomed.[7]

Considering the national level, the identification and definition of
national programmes for S&T development is usually a task for
government agencies, within the framework of a broad exercise of
S&T planning in which productive units and the scientific community should participate actively.[8]

From the perspective of international concerted financing, the national programmes that could receive support are those that comprise a broad range of S&T activities, including research, adaptation, technology importation, engineering design, quality control,
information, training, etc., related to critical problem areas whose
importance transcends national interests. These integrated S&T

development programmes would involve many types of institutions and could be the building blocks for international cooperation agreements. Large investment projects, particularly those financed by international sources, also deserve priority attention for international concerted S&T financing. It would be possible to use these large investment projects (power generation, irrigation schemes, steel mills, petrochemical complexes, automotive plants, etc.) as the nucleus around which to develop endogenous S&T capabilities. There are examples in this regard, ranging from the construction of atomic power plants in Argentina, to the development of a steel mill in Egypt, and to the establishment of a petrochemical complex in South Korea.

Within the framework of cooperative schemes for S&T development, it would be possible to establish subregional and regional programmes, defining areas of concentration to be supported during a specified period and providing a context for the submission and approval of proposals from member-country institutions. This would combine a central orientation with individual initiative. A mechanism of this type has been put into effect for the Special Projects Fund of the Regional Programme for Scientific and Technological Development of the Organization of American States. The Andean Pact's Andean Technology Development Projects system involves greater centralization, but also works in a similar fashion.[9]

In order to avoid excessive rigidity and to maintain flexibility, any subregional or regional funding mechanisms should contemplate the possibility of responding to specific project initiatives from two or more member countries, allocating matching funds to them, even if they fall out of the overall priority areas defined at a regional or subregional level. In addition, priority should be given to financing projects that involve technology transfer between countries in the region, particularly when consulting and engineering firms are involved.

With regard to global and interregional projects, it should be possible to establish mechanisms, such as the Consultative Group on International Agricultural Research and the International Research Programme on Tropical Diseases, in order to define priorities for problems of general concern. The two mechanisms mentioned, and many others of a similar nature, have led to the definition of lines of research and to the establishment of international training programmes.[10]

There is also a need to expand existing and to establish new global and interregional programmes in areas that have not received priority international attention. Among these it is possible to highlight the recovery and upgrading of traditional technologies, the improvement of technology importation by Third World countries, the absorption and assimilation of imported technology, the provision of risk capital for innovations originating in Third World countries, and the expansion and improvement of testing, standardization, and quality control systems. In this regard, it is preferable to create many relatively small and flexible organizations that could provide support to the initiatives and needs of the developing countries (witness the example of the International Foundation for Science in the field of basic research).

Finally, there is the need for expanding the physical infrastructure for endogenous S&T development (laboratories, research centres), and in this regard a world programme could be organized, focusing on the needs of the least developed countries, but always linking the development of physical infrastructure to programmes and projects. Similar remarks apply to the expansion of the human resources base.

As a concluding note, it is important to point out the role that nongovernmental organizations, and the scientific and professional associations in particular, have to play in the identification and definition of projects for international concerted financing. The world scientific community should participate actively in the process of mobilizing international funds for developing endogenous S&T capabilities in the Third World.[11]

B. Financing Mechanisms at the National Level

Accepting that financing the development of endogenous S&T capabilities is primarily a national responsibility, the first obvious task is to mobilize local resources to raise the total level of investment in S&T in the Third World. In 1974 the level of allocations for R&D (which is the figure quoted in most international statistics) was around 0.33 percent of GNP.[12] Developing countries should make efforts to reach at least 0.7 percent of their GDP, or even 1.0 percent, using a variety of means to reach these targets within the next decade.

A first mechanism would consist of expanding budgetary allocations for S&T. A manageable target from the point of view of government involvement in S&T financing would require specifying a percentage of government expenditures that should be devoted to S&T. Table 4.1 contains estimates that indicate the

TABLE 4.1(a) Gross Domestic Product (GDP), Government Expenditures, and Percentage Allocation for Science and Technology, 1976 (or most recent year)

CARIBBEAN AND LATIN AMERICA

Country	GDP (in US$ million)	% of GDP for S & T 0.7%	% of GDP for S & T 1%	Central Govt. Expenditure (in US$ million)	% of Govt. Expenditure in S & T 2%	% of Govt. Expenditure in S & T 3%
Antigua	22	0.15	0.22	--	--	--
Argentina	49106	343.74	491.06	491	9.82	14.73
Barbados	379	2.65	3.79	72	1.44	2.16
Belize	107	0.75	1.07	14	0.28	0.42
Bolivia	2154	15.08	21.54	280	5.60	8.40
Brazil	144615	1012.31	1446.15	14462	289.24	433.86
Chile	8088	56.62	80.88	1132	22.64	33.96
Colombia	13574	95.02	135.74	1086	21.72	32.58
Costa Rica	2345	16.41	23.45	352	7.04	10.56
Dominican Rep.	3915	27.40	39.15	313	6.26	9.39
Ecuador	4955	34.68	49.55	496	9.92	14.88
El Salvador	2186	15.30	21.86	262	5.24	7.86
Grenada	41	0.29	0.41	--	--	--
Guatemala	4363	30.54	43.63	324	6.48	9.72
Guyana	438	3.07	4.38	131	2.62	3.93
Haiti	1166	8.16	11.66	--	--	--
Honduras	1201	8.41	12.01	168	3.36	5.04
Jamaica	3045	21.31	30.45	548	10.96	16.44
Mexico	79139	553.97	791.39	18993	379.86	569.79
Nicaragua	1835	12.84	18.35	165	3.30	4.95
Panama	2028	14.20	20.28	304	6.08	9.12
Paraguay	1699	11.89	16.99	102	2.04	3.06
Peru	9872	69.10	98.72	987	19.74	29.61
Surinam	502	3.51	5.02	80	1.60	2.40
Trinidad/Tobago	2379	16.65	23.79	285	5.70	8.55
Uruguay	3693	25.85	36.93	480	9.60	14.40
Venezuela	31019	217.10	310.19	4653	93.06	139.59
TOTAL	373866	2617.00	3738.66	46180	923.60	1385.40

TABLE 4.1(b)   Asia and the Middle East

Country	GDP (in US$ million)	% of GDP for S & T		Central Govt. Expenditure (in US$ million)	% of Govt. Expenditure in S & T	
		0.7%	1%		2%	3%
Bahrain	244	1.70	2.44	117	2.34	3.51
Cyprus	778	5.45	7.78	14023	280.46	420.69
Iran	66777	467.44	667.77	2718	54.36	81.54
Iraq	13589	95.12	135.89	5183	103.66	155.49
Israel	13640	95.50	136.40	458	9.16	13.74
Jordan	1206	8.44	12.06	1809	36.18	54.27
Kuwait	11307	79.15	113.07	149	2.98	4.47
Lebanon	1488	10.42	14.88	942	18.84	28.26
Oman	2141	14.99	21.41	—	—	—
Qatar	159	1.11	1.59	—	—	—
Saudi Arabia	43924	307.47	439.24	7906	158.12	237.16
Syria	5904	41.33	59.04	1180	23.60	35.40
Turkey	41051	287.36	410.51	5747	114.94	172.41
Yemen	1113	7.79	11.13	111	2.22	3.33
Yemen Dem.	170	1.19	1.70	—	—	—
TOTAL	203491	1424.46	2034.91	40343	806.86	1210.29

TABLE 4.1(c) Asia -- East and Southeast (excluding Japan)

Country	GDP (in US$ million)	% of GDP for S & T		Central Govt. Expenditure (in US$ million)	% of Govt. Expenditure in S & T	
		0.7%	1%		2%	3%
Afghanistan	2207	15.45	22.07	--	--	--
Bangladesh	6838	47.87	68.38	132	2.64	3.96
Bhutan	48	0.34	0.48	--	--	--
Brunei	179	1.25	1.79	--	--	--
Burma	3474	24.32	34.74	--	--	--
Kampuchea	1157	8.10	11.57	198	3.96	5.94
Hong Kong	9322	65.25	93.22	652	13.04	19.56
India	86152	503.06	861.52	8615	172.30	258.45
Indonesia	37269	260.88	372.69	3727	74.54	111.81
Korea, Rep. of	25369	177.58	253.69	3044	60.88	91.32
Laos	203	1.42	2.03	--	--	--
Malaysia	11020	77.14	110.20	1984	39.68	59.52
Maldives	10	0.07	0.10	--	--	--
Nepal	1340	9.38	13.40	--	--	--
Pakistan	14510	101.57	145.10	1596	31.92	47.88
Philippines	17795	124.56	177.95	1776	35.52	53.28
Singapore	5915	41.40	59.15	651	13.02	19.53
Sri Lanka	3131	21.92	31.31	344	6.88	10.32
Thailand	16283	113.98	162.83	1786	35.72	53.58
TOTAL	242222	1695.54	2422.22	24505	490.10	735.15

TABLE 4.1(d) Africa (countries for which government expenditure is available)

144

Country	GDP (in US$ million)	% of GDP for S & T 0.7%	S & T 1%	Central Govt. Expenditure (in US$ million)	% of Govt. Expenditure in S & T 2%	3%
Algeria	16497	115.48	164.97	2310	46.20	69.30
Benin	1645	11.51	16.45	313	6.26	9.39
Botswana	83	0.58	0.83	20	0.40	0.60
Burundi	273	1.91	2.73	19	0.38	0.57
Central Afr. Rep.	331	2.32	3.31	73	1.46	2.19
Chad	391	2.74	3.91	74	1.48	2.22
Egypt	14905	104.33	149.05	3726	74.52	111.78
Eq. Guinea	76	0.53	0.76	8	0.16	0.24
Ethiopia	2648	18.54	26.48	344	6.88	10.32
Gabon	2158	15.10	21.58	259	5.18	7.77
Ghana	4052	28.37	40.52	486	9.72	14.58
Ivory Coast	4672	32.70	46.72	794	15.88	23.82
Kenya	3405	23.83	34.05	579	11.58	17.37
Lesotho	140	0.99	1.40	14	0.28	0.42
Liberia	927	6.49	9.27	93	1.86	2.79
Libya	13164	92.15	131.64	3554	71.08	106.62
Madagascar	1550	10.85	15.50	248	4.96	7.44
Malawi	723	5.06	7.23	94	1.88	2.82
Mali	403	2.82	4.03	64	1.28	1.92
Mauritania	191	1.34	1.91	31	0.62	0.93
Mauritius	588	4.12	5.88	76	1.52	2.28
Morocco	8083	56.58	80.83	1299	25.98	38.97
Niger	400	2.80	4.00	52	1.04	1.56
Nigeria	25120	175.84	251.20	3266	65.32	97.98
Reunion	1251	8.76	12.51	200	4.00	6.00
Rwanda	330	2.31	3.30	40	0.80	1.20
Sno Tome and Prin.	28	0.20	0.28	4	0.08	0.12

TABLE 4.1(d) Cont.

Country	GDP (in US$ million)	% of GDP for S & T 0.7%	1%	Central Govt. Expenditure (in US$ million)	% of Govt. Expenditure in S % T 2%	3%
Sierra Leone	625	4.38	6.25	63	1.26	1.89
Sudan	4341	30.39	43.41	608	12.16	18.24
Swaziland	113	0.79	1.13	14	0.28	0.42
Togo	264	1.85	2.64	21	0.42	0.63
Tunisia	4450	31.15	44.50	667	13.34	20.01
Cameroon	2009	14.06	20.09	281	5.62	8.43
Tanzania	2686	18.80	26.86	403	8.06	12.09
Upper Volta	456	3.19	4.56	46	0.92	1.38
Zaire	3695	25.87	36.95	924	18.48	27.72
Zambia	2687	18.81	26.87	672	13.44	20.16
TOTAL	125360	877.54	1253.60	21739	434.78	652.17

Source: Yearbook of National Accounts Statistics 1977, Vol. II. United Nations 1978 (ST/ESA/STAT/SER. 0/7/Add.1).

145

amount of resources that would be allocated if the 0.7 and 1.0 percent of GDP targets were achieved and also gives estimates for allocations equivalent to 2 and 3 percent of government expenditures.

With regard to the rationalization of financial allocations for S&T, there is also a need for integrating government expenditures on S&T that are usually made under different and dispersed headings of the national budget. The idea would be to consolidate the various government S&T expenditures into a "national science and technology budget," such as has been tried by the Colombian government with assistance from UNESCO. This system would help in identifying gaps in government allocations for S&T. Even though the establishment of consolidated budgeting procedures cannot be expected by itself to lead to increased financing, the fact that gaps are more easily identified could give rise to supplementary allocations.

In addition to budgetary provisions, there are several possible schemes that could be put into practice for increasing the flow of resources to S&T activities at the national level in order to reach, and even exceed, the targets of 0.7 and 1.0 percent of GDP. Some of them have already been tried over several years and operate with reasonable success; others are new proposals that need to be explored in further detail. A brief description of these mechanisms follows.

- *Cess on a percentage of net income before taxes.* This approach, known as the "ITINTEC System," was put into effect by the Peruvian Government in 1970. In the case of the Industrial Technology Institute (ITINTEC), all industrial enterprises in Peru, regardless of their ownership, must set aside 2 percent of net income before taxes for the performance of technological research. On the basis of rough estimates for various regions of the Third World, around US$552 million per year could be generated for the Caribbean and Latin America, US$96.4 million for the Middle East, US$61.3 million for Africa, and US$243.3 million for East and South-East Asia, as indicated in Table 4.2. Appendix A contains a description of the basic features of the ITINTEC system. Other countries are now considering the establishment of similar systems, and the Indian Industries Development and Regulation Act contains provisions enabling the government to establish such a cess.[13]

TABLE 4.2 Gross Profit of Manufacturing Industries and Size of the 2% Cess Fund, 1976 (in millions of U.S. dollars)

	GDP	Manufacturing Share of GDPa		30%b as Gross Profits	2% of Gross Profits for R & D
		% of GDP	Value		
Caribbean and Latin America	387400	23.8	92201	27660	553.2
Middle East	189300	8.5	16091	4827	96.5
Africa, excluding South Africa	137700	7.4	10190	3057	61.1
East and Southeast Asia, excluding Japan	263100	15.4	40517	12155	243.1

aCountries for which data are not available are not included.

bThe percentage of gross profits with respect to the manufacturing share of GDP was taken for Colombia during several years, observing that it remained more or less at the 30 percent level. The same percentage was considered for other countries.

Source: Same as in Table 4.1.

- *Allocating for R&D the same amount as royalty payments.* This approach, followed in South Korea for several years, stipulates that industrial enterprises have to set aside an amount for the performance of S&T activities equivalent to their royalty payments.[14] On the basis of payments made during various years by selected Latin American countries for which data are available, Table 4.3 indicates that more than US$500 million could be generated annually through this mechanism.

- *Percentage of direct foreign investment flows.* This approach has not been made law in any country, but there are many agreements between governments and foreign investors that stipulate the latter's contribution toward scientific and technological activities. For example, an agreement between the Peruvian government and the Swedish firm Volvo specified that the latter would set up a testing and quality control laboratory as part of its investment in an engine manufacturing plant. Assuming as a rough estimate that 1 percent of foreign investment flows would be allocated annually to S&T, Table 4.4 indicates that around US$95 million could have been generated in Latin American countries between 1970 and 1975.

- *Percentage of sales by transnational corporations.* Another mechanism could be established specifying that subsidiaries

TABLE 4.3 Payment of Royalties and Fees by Selected Latin American Countries (latest available year)

Country	Year	Payments of Royalties and Fees (millions of US dollars)
Argentina[a]	1974	101
Brazil[a]	1976	272
Chile[a]	1972	17
Colombia[b]	1974	21
Mexico[b]	1974	336
Trinidad and Tobago[a]	1975	18

Sources: (a) United Nations Commission on Transnational Corporations, Transnational Corporations in World Development: A Re-examination, New York E/C.10/38, 20 March 1978, Table III-68. (b) Oswaldo Parra "Balanza de Pagos Tecnológica" Ciencia, Tecnología y Desarrollo Vol. 2, No. 4, Octubre-Diciembre 1978, pp. 397-520, cuadro No. 2.

of transnational corporations should spend on S&T activities in the host country the same percentage of sales as they spend in their home country. If subsidiaries of U.S. transnational corporations in Latin America were considered, and the average for all U.S. industry in 1976 were applied (R&D expenditures represented 1.9 percent of total sales) it would be possible to generate approximately US$500 million, as indicated in Table 4.5. The figures for other regions are lower, but would be considerably higher if sales of European-owned foreign affiliates were considered (see Table 4.6).

- *Percentage of loans given by development banks.* This mechanism, tried in Argentina by the National Industrial

TABLE 4.4   Latin America and the Caribbean: Direct Foreign Investment Flows and 1.0 Percent Allocation for Science and Technology

| | Foreign Investment (US$ millions) | | | 1.0 Percent for S & T | | |
	1970	1975	1970–1975[a]	1970	1975	1970–1975
Argentina	11.0	--	50.9	0.11	--	0.51
Brazil	196.0	1,006.5	3,826.0	1.96	10.07	38.26
Mexico	323.0	609.5	2,674.0	3.23	6.10	26.74
Bolivia	-75.9	53.4	-0.7	-0.76	0.53	-0.01
Colombia	43.0	40.1	209.3	0.43	0.40	2.09
Chile	-79.0	49.8	-657.8	-0.79	0.50	-6.58
Ecuador	88.6	189.3	649.4	0.89	1.89	6.49
Peru	-70.0	315.7	339.6	-0.70	3.16	3.40
Venezuela	-23.0	354.6	-261.3	-0.23	3.55	-2.61
Paraguay	3.8	14.2	57.9	0.04	0.14	0.58
Uruguay	--	--	--	--	--	--
Costa Rica	26.4	69.0	227.2	0.26	0.69	2.27
El Salvador	3.7	13.1	56.5	0.04	0.13	0.57
Guatemala	29.4	80.0	236.0	0.29	0.80	2.36
Honduras	8.4	10.4	34.6	0.08	0.10	0.35
Nicaragua	15.0	10.9	76.2	0.15	0.11	0.76
Haiti	2.8	2.7	28.0	0.03	0.03	2.80
Panama	33.4	10.2	148.8	0.33	0.10	1.49
Dominican Rep.	71.6	50.5	343.6	0.72	0.51	3.44
Barbados	8.6	22.9	74.0	0.09	0.23	0.74
Guyana	9.0	0.8	-33.8	0.09	0.08	-0.34
Jamaica	162.1	-1.8	536.3	1.62	-0.02	5.36
Trinidad/Tobago	63.2	191.3	614.4	0.63	1.91	6.14
TOTAL	851.1	3,093.1	9,229.1	8.51	31.05	94.81

[a]Cumulative values.

Source: CEPAL, on the basis of International Monetary Fund, Balance of Payments Yearbook.

TABLE 4.5   Sales of Manufacturing Subsidiaries of U.S. Transnational
Corporations in Selected Latin American Countries, 1976

Country	Total Sales (millions of US dollars)	1.9% of Sales for S & T
Argentina	2,194	41.69
Brazil	10,559	200.62
Mexico	6,557	124.58
Colombia	1,362	25.88
Chile	207	3.93
Peru	389	7.39
Venezuela	3,531	67.09
Panama	124	2.36
Countries of the Central American Common Market	562	10.68
Other Latin American countries	414	7.87
Caribbean countries	354	6.73
TOTAL	26,253	498.82

Sources: CEPAL, Tendencias y cambios en la inversión de las Empresas
Internacionales en los países en desarrollo y particularmente en América
Latina, Documento de Trabajo No. 12, CEPAL/CET, Septiembre 1978, cuadro
No. 23.
    The percentage of total sales represented by R & D expenditures was
obtained from Business Week, June 27, 1977, p. 84.

TABLE 4.6   Sales of Manufacturing Subsidiaries of U.S. Transnational
Corporations in Other Developing Regions, 1976

Region	Total Sales (millions of US dollars)	1.9% of Sales for S & T
Africa (excluding South Africa)	696	13.22
Middle East	369	7.01
Asia and Pacific (excluding Japan, Australia, and New Zealand	5,547	105.40
TOTAL	6,612	125.63

Source: U.S. Department of Commerce, Survey of Current Business, Vol. 58,
No. 3, March 1978.

Technology Institute (INTI) for more than a decade, involves setting aside a certain percentage of the loans given by development financing institutions to industrial enterprises. In the case of INTI, this percentage is 0.25, and the funds are provided as matching grants to industrial firms in order to set up technological centres in particular branches. A similar mechanism has been recently established in Brazil, where State financing organizations have to set aside 2 percent of their gross income to support S&T activities. Although figures would vary from country to country, depending on the relative weight of development financing agencies in total industrial financing, it is clear that contributions could be substantial.[15]

There are other procedures that have been tried to increase the allocations for S&T, such as the provision of tax incentives. However, the effectiveness of such incentives has been questioned in many instances, and there is a need for examining closely the impact of fiscal incentives for research and development. Preliminary studies show that their influence on increasing R&D expenditures is, at best, inconclusive.[16] Finally, there is also the possibility of setting up governmental and nongovernmental foundations, funds at the regional and state levels, project financing institutions, and so on, as has been done in Brazil, Mexico, India, and many other countries.

The main points to emphasize are that there are many possible national sources of funds for the development of endogenous S&T capabilities and that a variety of institutional and organizational arrangements is required. Furthermore, *all of these possible sources of funds can be established through national decisions and legislation, without the need for international agreements.* This is particularly important when dealing with S&T financing schemes that may affect subsidiaries of transnational corporations, and even more so when obligatory contributions related to the local operations of these corporations are contemplated. However, it might be possible for a group of Third World countries to agree on establishing similar systems at the national level and to coordinate their operation.

C. International Financing Mechanisms

International financing mechanisms should complement the resources allocated at the national level and, for the reasons mentioned earlier in this chapter, there should be a massive transfer of financial resources for endogenous S&T development from the industrialized to the Third World countries. There is a variety of schemes, some of which have been in operation for some time, to channel international financing to S&T in the developing countries. The resources transferred specifically for S&T development in the past have been relatively small, and it is necessary to devise and establish new mechanisms that would increase the resources substantially and that could be more effective in providing financing support for endogenous S&T development in the Third World.

First, there are the official bilateral assistance channels, which involve in most cases tied aid. The idea would be to provide at least 25 percent of all aid funds without any ties, allowing the recipient countries to make their own decisions with regard to the projects and purchases and, in general, with regard to the use of the funds. This would allow the recipient countries to "unpackage" the technology they receive in projects financed with official development assistance funds, improving the choice of technology and increasing decision-making autonomy on technological matters. Furthermore, 5 percent could be specifically earmarked for S&T development purposes; and if this was done, around US$800 million would be generated yearly, as indicated in Table 4.7. To channel these portions of bilateral development assistance earmarked for S&T development, special purpose institutions, such as the Canadian International Development Research Centre (IDRC) and the Swedish Agency for Research Cooperation with Developing Countries (SAREC), could be created. The United States and other Western industrialized countries are contemplating the establishment of similar agencies, although it is essential that they remain very flexible in their operating procedures to be of any assistance to the Third World. Appendix B contains a description of the essential characteristics of IDRC, which could be considered as a guide for similar institutional initiatives.[17]

A second channel for industrialized countries' support of endogenous S&T development in the Third World involves a variety of private agencies and institutions. In addition to the foundations that have been traditionally involved in supporting research and training (Ford Foundation, Rockefeller Foundation, Friedrich Ebert Foundation, Tinker Foundation, International Foundation for Science, etc.), it is necessary to explore new mechanisms in which transnational corporations and international banks could participate, but with the explicit objective of developing the *autonomous* S&T capabilities of Third World countries. Few constructive ideas have

TABLE 4.7 Official Development Assistance (ODA) from Industrialized to Developing Countries and Possible 5 Percent Allocation for Scientific and Technological Development, 1976 (in millions of U.S. dollars)

Country Development Assistance Committee (DAC) Members		Flows of Financial Assistance	5 Percent for Science and Technology
Sweden)	(countries attaining the 0.7%	607	30.35
Netherlands)	ODA target in 1976)	718	35.90
Norway)		215	10.75
France)		2,152	107.60
Denmark)		214	10.70
Belgium)	(countries above the DAC	336	16.80
Canada)	ODA average in 1976)	885	44.25
New Zealand)		53	2.65
Australia)		385	19.25
United Kingdom)		827	41.35
Germany, F. R.)		1,309	65.45
United States)		4,304	215.20
Japan)	(countries below the DAC	1,093	54.65
Switzerland)	ODA average in 1976)	111	5.55
Finland)		51	2.55
Italy)		215	10.75
Austria)		37	1.85
DAC TOTAL		13,512	675.60
Socialist countries of Eastern Europe (preliminary data):			
Bulgaria		8	0.40
Czechoslovakia		1,064	53.20
Germany, Dem. Rep.		105	5.25
Hungary		20	1.00
Poland		52	2.60
Romania		261	13.05
USSR		1,208	60.40
Socialist Countries TOTAL		2,718	135.90
Industrialized Countries TOTAL		16,230	811.50

Source: United Nations General Assembly. Committee Established Under General Assembly Resolution 32/174 (First Session, 3-12 May 1978, Agenda Item 2), Transfer of Resources in Real Terms to Developing Countries. Note by the Secretariat (Prepared by the Secretariat of the United Nations Conference on Trade and Development), 13 April 1978, Document Number A/AC.191/7--Tables 2 and 5.

been put forward in this area, and transnational corporations—which have been so innovative in many other areas—should certainly exert ingenuity and creativity in this field (not that we are very optimistic about this!).

A third channel to finance the development of S&T capabilities in the Third World involves the multilateral financial agencies that have been established during the last thirty years. In addition to the United Nations agencies and the United Nations Development Programme, there are the World Bank, the regional development banks, and the regional organizations, such as the U.N. Regional Economic Commissions, the Latin American Economic System (SELA), the Organization of American States, the Andean Pact, the Association of South-East Asian Nations (ASEAN), the Arab Education, Culture and Science Organization (ALECSO), and the Organization of African Unity (OAU). In view of the activities of multilateral financial institutions, 10 percent of all their loans could be specifically allocated to the development of endogenous S&T capabilities in Third World countries, encompassing the range of activities that was described earlier in this chapter. As Table 4.8 indicates, this could generate approximately US$450 million per year. Furthermore, it might be possible to establish a series of regional funds for S&T development, which would receive contributions from a variety of sources. Finally, in addition to the regional funds, a series of funding consortia could be established, along the lines of the Consultative Groups on International Agricultural Research and the International Programme of Research on Tropical Diseases, to obtain contributions from a variety of sources and multilateral agencies in particular.

There is also a need to establish new mechanisms that would generate additional funds for endogenous S&T development in a direct, automatic, and continuous way. Although the political feasibility of these new mechanisms may be open to debate, it is clear that they could play the major role in providing financial support. Among the proposals that have been made is the "development tax" scheme, which would levy a tax on domestic consumption, on traded commodities, or on incomes in the developed countries. The development tax could also be based on internationally traded goods or on the exploitation of mineral resources of the seabed. There have also been suggestions to link the generation of funds for development to the creation of Special Drawing Rights (SDRs) or to the profits made from gold sales by the International Monetary Fund. A certain percentage of the funds generated by such schemes could be

TABLE 4.8 Net Flow of Resources from Multilateral Institutions to Developing Countries in Africa, Asia, and Latin America and Possible Allocations for S & T Development, Including Soft Loans, Grants, and Hard Loans, 1976 (in millions of U.S. dollars)

	Flow of Resources		10% for S & T	
	(a)	(b)	(a)	(b)
African Development Bank	21	29	2.1	2.9
Asian Development Bank	291	269	29.1	26.9
European Development Fund	451	451	45.1	45.1
European Investment Bank	10	10	1.0	1.0
World Bank	890	1,254	89.0	125.4
International Development Association	1,279	1,186	127.9	118.6
International Finance Corporation	105	119	10.5	11.9
Interamerican Development Bank	156	203	15.6	20.3
United Nations agencies	1,009	993	100.9	99.3
TOTAL	4,212	4,514	421.2	451.4

Notes: (a) Figures in this column include transactions of multilateral institutions with all developing countries.

(b) Figures in this column exclude transactions of the multilateral institutions with the following oil-exporting countries: Algeria, Bahrain, Brunei, Ecuador, Gabon, Indonesia, Iran, Iraq, Kuwait, Libya, Nigeria, Oman, Qatar, Saudi Arabia, Trinidad and Tobago, United Arab Emirates, and Venezuela.

Source: UNCTAD Secretariat, Transfer of Resources in Real Terms to Developing Countries, A/AC. 191/7, 13 April 1978, Table No. 7.

channeled to the development of endogenous S&T capabilities in the Third World.[18]

Another innovative proposal is that made by Bernard Lietaer to establish a new North-South trading mechanism that would give stability to commodity prices and compensate Third World countries for losses due to inflation through a "World Development Exchange." Lietaer's proposals amount to a transfer of resources from industrialized to developing countries by means of a voluntary reversal of the deterioration of the terms of trade between commodities and manufactured products. As Lietaer's proposals are linked to investments in development projects, it would be possible to add a component of scientific and technological development to his trading mechanism and World Development Exchange.[19]

There have also been proposals made to levy a tax on military and armament expenditures by the industrialized nations. In 1975 world military expenditures were around US$350 billion, and even a small fraction of these expenditures could generate a very large amount of funds for development in general and for S&T in particular. If the figures for military R&D were considered, 5 percent of the US$30 billion spent in 1977 would have generated an additional US$1.5 billion for S&T development in the Third World. However, the difficulties in assessing military expenditures and the complications involved in collecting such a tax make this proposal one of the most difficult to define in operational terms.[20]

The Group of 77 proposed at the United Nations Conference on Science and Technology for Development (UNCSTD) the establishment of a fund linked to the imbalance in the trade of technology-intensive goods between industrialized and Third World countries. The rationale of the proposal is the fact that an objective indicator of technological disparities between North and South is the imbalance in the trade of manufactured goods of a certain degree of complexity, which incorporate the results of S&T research. Consequently, in order to initiate a gradual process of redistribution of the world scientific and technological effort within the framework of more equitable world trade patterns, it is appropriate to link the transfer of resources from developed to developing countries for the development of S&T capabilities to the imbalances in trade of technology-intensive products. Because of its importance and implications at UNCSTD, this proposal will be examined in a separate section below.

The proposals mentioned in this section have concentrated on the generation of international financial resources, but it is also necessary to identify the types of activities to which funds could be applied. However, the definition of activities to be financed with international resources should not be confused with requests for detailed and very specific project proposals that could delay indefinitely the allocation of funds. What is required is the identification of broad types of activities, not necessarily of problem areas, that would provide a framework for the initiatives of various national, subregional, regional, and international organizations, allowing for the possibility of the projects' and programmes' being redefined, focused, and modified during their execution. In this regard, it is possible to offer an illustrative list of the types of S&T activities that could be financed with international funds:

- A world programme for the development of the S&T infrastructure of the Third World, including the establishment of research centres, laboratories, and other physical facilities, as well as the expansion of the human resources base, including the establishment of training centres, higher education institutions, and cooperative programmes for graduate education in S&T.

- A special assistance programme for the least developed countries including, in addition to the development of an infrastructure for S&T, the transfer of technology in preferential terms and the support for the adaptation, absorption, and diffusion of imported technology.

- Regional and subregional programmes based on common problems, covering the whole spectrum of activities from basic research to the establishment of pilot and demonstration plants. These would be the subregional and regional equivalents of the Integrated Programmes for Technological Development mentioned earlier for the national level.

- Regional and subregional programmes focusing on the transfer and adaptation of technology, and particularly on the transfer of technology among developing countries. This would include the search for technologies, information exchange, training, disaggregation, adaptation, and other activities linked to the importation and absorption of technology.

- Regional and subregional programmes to develop S&T services, including S&T information and documentation, quality control and standards, registries of licensing agreements, etc.

This is just an illustrative list, and it is clear that a major effort would be required to identify and select those activities linked to the development of endogenous S&T capabilities that should receive international concerted financing, differentiating them from those that are primarily a national responsibility. In this regard, the Vienna Programme of Action agreed upon at UNCSTD in 1979 provides guidelines for the identification of priorities.

4. The Group of 77 Proposals at UNCSTD

A. Background to the Group of 77 Proposals

Among the various schemes advanced to establish international financing mechanisms for science and technology, the proposals made by the Group of 77 at UNCSTD are of particular importance, primarily because of the general support given by the developing countries to them and the (albeit reluctant) acceptance by the industrially developed countries of the need to establish an international financing mechanism. This acceptance was expressed through the willingness of the industrialized countries to set up a two-year Interim Fund as a forerunner of the financing system and by the agreement to convene an intergovernmental group of experts to design the latter.

Although the real goodwill and political support for these initiatives are still to be tested (and the indications are that the Interim Fund and the financing system will not receive significant support from the key industrialized nations), the agreements reached in Vienna in August 1979 at UNCSTD constitute the basis for an initiative that could, eventually, lead to the establishment of an international financing mechanism for promoting the development of science and technology in the Third World.

The new financing system is scheduled to begin operating on 1 January, 1982, and the intergovernmental group of experts has to complete the work of defining the nature and operational characteristics of the system before then. During the same period the Interim Fund, to be sustained by voluntary contributions, will begin to support activities related to the Vienna Programme of Action.

It is important to indicate that the Group of 77 did not propose the creation of the Interim Fund. Their efforts were directed towards the establishment of the proposed "United Nations Financing System for Science and Technology for Development" (the capital letters in "Financing System" were abandoned during the negotiations in Vienna!). The idea of setting up the Interim Fund was formally introduced by the U.S. delegation and was agreed to, with the support of all delegations at UNCSTD, as part of a package negotiation and compromise on institutional and financial matters.

The main ideas regarding financial matters in the Group of 77 position at the Vienna Conference emerged out of the work of a few delegations and institutions from the developing countries during 1978 and 1979. The first proposals to be put on the table, those referring to the "trade imbalance fund" that ultimately became the Group of 77 position, were made at a meeting convened in Caracas in May 1978 by the Venezuelan Council of Science and Technology (CONICIT) and the Latin American Economic System (SELA).[21] They were subsequently adopted by the Venezuelan government and proposed formally at a meeting of the Andean Group in July 1978. The Andean Group secretariat enriched and developed further the initial proposals, and they became the official position of the Andean countries at the Latin American regional meetings in preparation for UNCSTD. An ad hoc working group was created by the UN Economic Commission for Latin America to discuss these issues, and the report prepared by this working group after its March 1979 meeting became the most complete statement made by a group of developing countries on financial matters during the preparatory process for the Vienna Conference.

In early 1979, when the Latin American proposals began to be discussed within the Group of 77, other delegations presented financial proposals of their own. The Yugoslavian delegation presented a rather complex scheme to establish a financing system; the Jamaican delegation proposed the creation of a new fund, to be administered by UNDP and sustained by the proceedings from disarmament measures; the Mexican delegation suggested alternative ways of calculating assessed contributions; and several African countries made a number of proposals regarding the organization and structure of international financing schemes. All these proposals were discussed in detail and many of their components were incorporated into the final Group of 77 position, even though the views of the Latin American delegations (and those of the Andean Group in particular) were the main influence on the final Group of 77 position on financial matters. These proposals crystallized at the

time of the fifth meeting of the UNCSTD Preparatory Committee in
June 1979 and were discussed thoroughly at the technical level at
the Schloss Hernstein meeting convened by UNITAR in July 1979.

The negotiation process at the Vienna Conference focused with
particular intensity on institutional and financial issues. In the end,
agreement was reached around the compromise proposed by the
U.S. delegation, for which some groundwork had been done by one
of the UN agencies and a few delegations from key industrially
developed and developing countries.

B. Basic features of the Group of 77 Financing System

The main principles on which the Group of 77 based its position
on financial matters at the conference in Vienna can be summarized
as follows:[22]

1. That it was necessary to establish a new financing system for
   science and technology for development, independent of existing financing schemes, and clearly identified as a separate
   entity;

2. That the resources allocated to this new financing system
   should be additional to those already existing within and outside the United Nations for similar purposes;

3. That the funds would be used specifically for strengthening
   endogenous S&T capabilities through a variety of capacity-
   building activities;

4. That the resources should be complementary to the national
   efforts of developing countries;

5. That contributions to the financing system should be based
   on considerations regarding the asymmetry of S&T capabilities between industrially developed and developing
   countries, that they should be predictable, assured, and continuous, and that they should be untied;

6. That the financing system should express the solidarity of the
   developing countries and provide special treatment to the
   least developed among them;

7. That there should be both assessed and voluntary contributions to the financing system;

8. That the financing system should be under the substantive
   control of the developing countries; and

9. That the administrative machinery should be light and should not impose a heavy burden on the resources of the financing system.

These principles guided the Group of 77 position during the negotiation process at UNCSTD, and presumably they will guide the participation of the developing-country members of the intergovernmental group of experts that will design the new financing system. All the basic ideas behind these principles have been practically adopted in the Vienna Programme of Action and, if the developed countries have the political will to back them up, these principles could lead to the establishment of an innovative and effective international financing mechanism for science and technology for development.

One of the key issues discussed at length during the preparation of the Group of 77 proposals on financial matters referred to the objectives of the system. The Vienna Programme of Action states clearly the range of activities to be financed and defines the general objective of the financing system, namely the development of endogenous scientific and technological capabilities in the developing countries, but there still remain some questions regarding the differentiation of programmes and projects to be supported by the new financing system from those financed by existing UN sources, particularly in UNDP, and related in some way or another to science and technology.

The specificity of the new financing system can be appreciated more clearly if a difference is established between programmes and projects aimed primarily at the building of endogenous S&T capabilities and those aimed at solving specific development problems through the use of science and technology. The first corresponds to what has been called the "capacity-building approach" and the second to the "problem-solving approach."

In the capacity-building approach emphasis is placed on the development of endogenous scientific and technological capabilities, linking the growth of science with the evolution of modern technology and the selective recovery of the traditional technological base and integrating science and technology into productive and service activities.[23] The main idea is to treat the question of science and technology for development in a "horizontal way," giving priority to capacity building across the board. Issues such as self-reliance and autonomy of decision making in science and technology matters figure prominently in this approach.

The problem-solving approach is characterized by the emphasis it places on the solution of specific technical problems, the identification of a limited number of priority areas, and the selection of areas of concentration, seeking to develop science and technology in a "vertical way" and leaving aside aspects such as science and technology policies, international cooperation, regulation of technology imports, and so on.

However, the difference between projects and programmes aimed at building endogenous S&T capabilities and those aimed at solving specific problems through the use of science and technology is one of degree, and the interrelations between these two aims need to be clarified in the context of specific programmes and projects. Nevertheless, in general terms, a project aimed primarily at capacity building could be defined as one in which the training, establishment of infrastructure, support of autonomous decision making, and creation of the right context for S&T development have at least as much weight in the project objectives as the finding of a solution to a specific problem through S&T activities. Although it is acknowledged that the problem-solving process itself may also contribute to the building of S&T capabilities, it is clear that not all S&T activities directed towards solving a specific problem of development would contribute to capacity building. At one extreme, and accepting that most of the S&T research capacity is concentrated in the highly industrialized countries, a number of problems of development could be solved in the industrially developed countries with no impact whatsoever on the development of endogenous S&T capabilities in the Third World.

The differences and complementarities between a problem-solving and a capacity-building approach to the support of science and technology for development and, in particular, to the identification, formulation, evaluation, and execution of programmes and projects can be appreciated better through the examination of some characteristics associated with the execution of a specific project. The first is the time it would take to solve a problem from a strictly S&T point of view. If capacity-building considerations were introduced into a project (training, participation of a larger number of local institutions, decision-making autonomy, and possibility of learning while doing, etc.) it is likely that the project would take longer than if no capacity-building considerations had been introduced.

The risks involved in the execution of a given project would also reflect an emphasis on problem solving or capacity building. Greater risks would be associated with a capacity-building approach because of the relative lack of experience of the participating institutions and individuals, the need to train junior researchers, and the possible inadequacy of facilities. Similarly, the cost of the project would also be affected, for the greater the capacity-building component introduced, the more costly the project is likely to be.

Furthermore, the concept of capacity building will mean different things to countries at different stages of S&T development. For those with little or no capabilities in science and technology it will involve establishing the basic educational and institutional infrastructure; for those with some S&T capacity it will imply a linkage of their S&T infrastructure with the productive system, and for those relatively more advanced, it may imply developing an aggressive research and innovation capacity to attain world leadership in some fields of science and technology.

The explicit consideration of a capacity-building dimension in the operation of the financing system proposed by the Group of 77 would require a substantive revision and modification of the project formulation, evaluation, and monitoring procedures that characterize the regular operations of most international funding agencies. Returns on investment in capacity building are seen only in the long run, are diffused through the whole productive and social structure (many externalities), and involve greater risk than projects oriented exclusively towards problem solving. These features would require new methods of project appraisal and evaluation.

Another feature of the Group of 77 proposals refers to the quantitative targets for the financing system, which were placed at the level of $2 billion for 1985 and $4 billion for 1990. These targets were arrived at through different calculations:

a. The $2 billion sum was equivalent to one-fourth of the 1 percent of the developing countries' GNP that the World Plan of Action of the UN Advisory Committee on Science and Technology for Development recommended should be devoted to science and technology for development, and the idea was that the financing system should contribute about one-quarter of the total efforts of developing countries to reach the 1 per cent target;

b. These targets also correspond to the target of 0.05 percent of their GNP that the industrially developed countries, as part of the International Development Strategy, had agreed to contribute directly to the development of S&T in the developing countries;

c. Finally, 2 percent of the average trade imbalance in manufactured goods between industrially developed and developing countries in the mid-1970s also amounted to about $2 billion. Indeed, these calculations led to the figure of $2 billion for 1980, but it was felt that a shift in time would make the proposal more palatable to the developed countries, particularly when comparing it to total Official Development Assistance flows of the OECD countries, which are expected to exceed $40 billion by 1985.

Furthermore, considering that the R&D expenditure of developing countries in 1973 was $2.8 billion, it became apparent that the targets of $2 billion for 1985 and $4 billion for 1990 would be well within the absorptive capacity of developing countries, but would still have a significant impact on the development of endogenous S&T capabilities.

Another key feature of the Group of 77 proposals on financial matters refers to the method of calculation of the contributions to the proposed financing system. The Group of 77 linked contributions to the fund to the trade imbalances in manufactured goods between industrially developed and developing countries. There were many reasons for making this specific proposal, some of which were related to the need to use an easily agreed-upon statistical base, such as international trade figures. However, the basic justification for linking the contributions to the trade imbalances in manufactured goods was that these imbalances reflect technological superiority and dependency.

The theoretical background that justifies linking technology issues with trade performance emerged during the last twenty years as a result of many studies that showed the interactions between these two factors. For example, Posner, Hufbauer, Vernon, Hirsch, and Gruber, Metha, and Vernon, among others, advanced a variety of theoretical and empirical arguments linking technological advantages with superior trade performance.[24] The OECD analytical report, *Gaps in Technology*, devoted a full chapter to this issue,[25] and a conference of the National Bureau of Economic Research in the United States examined the relations between technology and international trade.[26] This concern has also been reflected in the work of economists, such as Johnson[27] and Stewart,[28] and has spawned several studies in the United States during recent years.[29]

Although opinion in academic circles is far from unanimous regarding the precise relationships between endogenous scientific and technological capabilities and trade performance, it is clear that a trade surplus in technology-intensive goods is associated with superior technological capabilities. Fajnzylber[30] has published a paper in which he highlighted the extent of these trade advantages linked to technology capabilities. He indicated that in 1977 the industrially developed market economies accounted for 87.7 percent of world exports in engineering products, centrally planned industrially developed economies accounted for 10 percent, and the developing countries for only 2.5 percent (see Table 4.9). Furthermore, between 1969 and 1976, the industrially developed market economies increased their surpluses in the trade of capital goods from $17 billion to $77 billion while the developing countries' deficit grew from $16 billion to $72 billion.

Considering trade in all manufactured products between Latin America and the United States, Japan, and the European Economic Community in 1975, figures indicate that the trade imbalance against Latin America was approximately US$23.3 billion. According to the Group of 77 proposals, if 2 percent of that imbalance had been contributed to the fund for S&T, US$460 billion would have been generated.[31] More detailed estimates, carried out by the Secretariat of the Andean Group, put the figures for Latin America at the level of US$217 million, considering the average trade im-

TABLE 4.9   Imports and Exports of Engineering Products, 1977 (in millions of U.S. dollars)

	Exports Value	%	Imports Value	%	Exports Imports (ratio)
Developed Countries:					
Market economies	273,585.5	87.5	183,844.6	58.8	1.49
Centrally planned economies	31,132.5	10.0	32,558.3	10.4	0.96
Developing Countries	7,782.0	2.5	93,178.0	29.8	0.08
Unaccounted Imports	--		2,919.1	0.9	
WORLD TOTAL	312,500.0	100	312,500.0	100	

Source: Economic Commission for Europe, Bulletin of Statistics on World Trade of Engineering Products, 1977, New York, 1979. Taken from F. Fajnzylber "Industrialización, bienes de capital y empleo en las economías avanzadas," Comercio Exterior, Vol. 30, No. 8, México, Agosto 1980, pp. 867–880.

balance between 1969 and 1973 and using the same 2 percent figure. The corresponding estimates for Asia and Africa for the period 1970–1974 lead to a contribution by the industrially developed countries of US$284 million and US$187 million, respectively. The Andean Pact estimates exclude trade in armaments and also introduce a redistribution factor to reach more equitable allocations among the developing countries in a given region.[32] Finally, a projection of trade imbalances between the United States, Japan, and the European Economic Community and all developing countries in two technology-intensive sectors has been carried out by Pizano and Perry. Their estimates for the total amount to be contributed, based on projections of the trade imbalance in chemical and engineering products and on the use of 2 percent as the level to determine contributions, lead to US$3.2 billion in 1980 and to US$4.2 billion in 1982.[33]

Therefore, the Group of 77 proposals to link contributions to the financing system for science and technology for development to the trade imbalances in manufactured goods appears eminently sensible. However, it is clear that modifications need to be introduced to make the method of calculation more equitable, particularly to the small industrially developed nations in which foreign trade accounts for a relatively high proportion of their Gross National Product, to clarify the nature of the traded goods on which to base the assessment of contributions, such as considering engineering products instead of manufactures, and also to introduce considerations regarding the levels of per capita income and gross domestic investment. Indeed, at the UNITAR Schloss Hernstein meeting, where the Group of 77 proposals were examined in July 1979, one of the participants (a highly respected economist and a senior executive at the World Bank), summarized the feelings of most of the participants from industrially developed and developing countries that none of the objections raised to the financing proposals could not be overcome and that the discussions had indeed highlighted the perceived advantages of the proposed financing system.

Conclusions

This chapter has examined the problems involved in the financing of endogenous scientific and technological development in the Third World and has reviewed a variety of proposals with the aim of stimulating reflection and discussion. The financial proposals made by the Group of 77 at UNCSTD in Vienna received particular attention, but it is clear that other proposals to establish national and regional mechanisms are equally important and that such ideas as the possibility of establishing a Third World funding mechanism also deserve to be explored.

From this perspective, the diversity of activities to be supported and of conditions prevailing in the developing countries indicates the need for a variety of financial channels. The decision to establish the UN financing system for science and technology for development should be implemented as soon as possible. Furthermore, even though the financing system should be a separate and clearly identifiable entity, it could draw on the expertise, organizational capabilities, and experience of existing international organizations, such as the United Nations Development Programme, the Regional Development Banks, the Latin American Economic System (SELA), the Organization for African Unity, the Association of Southeast Asian Nations (ASEAN), and the Andean Group.

Unfortunately, the experience with the Interim Fund during its first year does not appear too encouraging if we consider it as an indicator of the political will of industrialized nations to contribute to the development of science and technology in the Third World. Even though the Interim Fund has generated well over 600 proposals with requests for a total of more than US$400 million, the voluntary contributions to the fund have not even reached one-fifth of the agreed target of US$250 million.

In concluding, it is appropriate to say that the various financial mechanisms and schemes mentioned in this chapter indicate that a substantial increase in the resources allocated specifically for Third World endogenous scientific and technological development should be possible. The feasibility of many schemes has already been proved in practice, and there are several others that could be established with little difficulty. Many of the mechanisms do not require international agreements, and it would be possible to undertake concerted but autonomous actions in several Third World countries to establish national mechanisms of a similar type. Finally, the UN has agreed to establish a financing system for science and technology for development.

But even if all these mechanisms to finance the development of endogenous S&T capabilities in the Third World were put into effect simultaneously and at all levels through a miraculous exercise of coordinated political will, their impact would be felt only in the long run and would only just begin to alter the existing and cumulative disparities in S&T capabilities between industrialized and Third World countries. Rather than being a cause for cynicism and despair, this should be taken as a challenge to be met squarely by the Third World with or without the help of the industrially developed countries,' for there is no escape from the condition of underdevelopment unless endogenous S&T capabilities are acquired.

Notes

This chapter was written in April 1979 while the author was with the International Development Research Centre (Canada), on secondment to the Secretariat of the United Nations Conference on Science and Technology for Development, and was revised in August 1980.

1. Jan Annerstedt, *A Survey of World Research and Development Efforts* (Roskilde [Denmark]: Institute of Economics and Planning, Roskilde University Centre, July 1979).

2. Francisco Sagasti, "Towards Endogenous Science and Technology of Another Development," *Development Dialogue*, No. 1, 1979, pp. 13–23; and *Technology, Planning and Self-Reliant Development* (New York: Praeger Publishers, 1979).

3. See *Ways and Means of Accelerating the Transfer of Real Resources to Developing Countries on a Predictable, Assured and Continuous Basis*, Report of the Secretary General, A/31/186, 21 September 1976; *Transfer of Resources in Real Terms to Developing Countries*, note by the UNCTAD Secretariat, A/AC 191/7, 13 April 1978; and *Algunos aspectos de las perspectivas de las corrientes financieras globales en el contexto del Tercer Decenio de las Naciones Unidas para el Desarrollo*, Informe de la Secretaria de la UNCTAD, TD/B/C.3/161/Supp. 3, 20 de Junio 1980.

4. For a discussion of financial policy instruments to affect the pattern of demand for technology, see *Science and Technology for Development: Main Comparative Report of the STPI Project* (Ottawa: International Development Research Centre, 1978); and Francisco Sagasti, "El financiamiento industrial como instrumento de política tecnológica," *El Trimestre Económico*, Vol. 45, No. 178, Abril-Junio 1978, pp. 401–442. A new project on this subject is being launched by the OECD Development Centre, the Office for Science and Technology of the World Bank, and the International Development Research Centre, under the coordination of Nicolas Jéquier.

5. See the Latin American regional paper for the UN Conference on Science and Technology for Development (Document A/Conf.81/PC16/Add.1), 29 January 1979.

6. For a discussion of this issue, see the Swedish national paper for the UN Conference on Science and Technology for Development; the report of the Dag Hammarskjöld Seminar on Science and Technology for Develop-

ment, December 1978, which appeared in *Development Dialogue*, No. 1, 1979; and the *Report of the Jamaica Symposium on Mobilizing Technology for World Development* (Washington, D.C.: International Institute for Environment and Development, January 1979). The Brandt Commission Report, *North/South: A Programme for Survival* (Cambridge, Mass.: M.I.T. Press, 1980), also covers this issue at length.

7. The set of proposals submitted to the UN Interim Fund for Science and Technology for Development (see section 4), involving more than 600 projects by October 1981 and totaling more than US$400 million in requests, indicates that there is a capacity to articulate financing requests for S&T in many developing countries.

8. See Francisco Sagasti and Alberto Araoz (eds.), *Science and Technology for Development: Planning in the STPI Countries* (Ottawa: International Development Research Centre, 1979). For an annotated bibliography on the subject see Wolfgang Mostert, *La planificación de la ciencia y la tecnología en los países en desarrollo* (Lima: Escuela Superior de Administración de Negocios [ESAN], 1976). A comprehensive review of the subject is contained in Francisco Sagasti, "Science, Technology and Development Planning: A Review of Key Issues," in K.-H. Standke and Anandakrishnan (eds.), *Science, Technology and Society* (Oxford: Pergamon Press, 1980).

9. For descriptions of these programmes, see UN Economic Commission for Latin America (ECLA), *International Machinery for the Financing of Scientific and Technological Development*, E/CEPAL/L.189, 21 March 1979; and Junta del Acuerdo de Cartagena, *Technology Policies in the Andean Pact* (Ottawa: International Development Research Centre, 1976).

10. See Consultative Group on International Agricultural Research, *International Agricultural Research* (New York: World Bank/FAO/UNDP, 1976), and *Tropical Diseases* (Geneva: WHO/UNDP, no date).

11. See, for example, the statements and reports produced at the Singapore Symposium on Science and Technology for Development in January 1979, with the participation of more than 120 scientists from all parts of the world and the sponsorship of eighteen scientific and professional organizations.

12. UNESCO, *Estimation of Human and Financial Resources devoted to R&D at the World and Regional Level* (Paris, May 1979).

13. Government of India, The Industries (Development and Regulation) Act, 1951 (as modified up to 1 December 1975). In particular see Chapter 2, section 9.

14. The provision is contained in the South Korean Technology Promotion Law. For a discussion see *Final Report of the Korean STPI [Science and Technology Policy Instruments] Project* (Seoul: Korean Advanced Institute of Science, 1976).

15. For a description and critique of the Argentinian system of the Instituto Nacional de Tecnología Industrial (INTI) see O. Ozlak and M. Cavarozzi, *El INTI y el desarrollo tecnológico de Argentina*, report of the Argentinian STPI team (Buenos Aires, 1976).

16. For a discussion on some findings regarding the impact of fiscal incentives, see Alejandro Nadal, *Instrumentos de Política Científica y Tecnológica en México*, final report of the Mexican STPI team (Mexico City: El Colegio de México, 1976).

17. The International Development Research Centre (IDRC) of Canada and the Swedish Agency for Research Cooperation with Developing Countries (SAREC) have channeled slightly less than 5 percent of the bilateral assistance funds of their respective countries. In 1977/1978 the IDRC budget was Can.$38.3 million, and the SAREC budget for 1977/1978 was 91.4 million Swedish Crowns.

18. See the review of different mechanisms provided by Eleanor Steinberg and Joseph Yaeger, *New Means of Financing International Needs* (Washington, D.C.: Brookings Institution, 1978); and the *Study on Financing the United Nations Plan of Action to Combat Desertification*, prepared by a group of high-level specialists in international financing convened by the Executive Director of the United Nations Environment Programme pursuant to General Assembly Resolution 34/184, Geneva, 25 July 1980.

19. Bernard Lietaer, *A Role for Europe in the North-South Conflict* (Brussels: European Cooperation Fund, 1978).

20. The figures are all taken from Ruth Leger Sivard, *World Military and Social Expenditures, 1978* (Leesburg, Va.: WMSE Publication, 1978). She also added in this report: "The search for new and more destructive weapons dwarfs all other research efforts, whether publicly or privately financed. Weapons research occupies over half a million scientists and engineers throughout the world and takes more public research money than all research on energy, health, education, food and other civilian needs combined" (p. 9).

21. For the initial ideas related to the "trade imbalance fund" see Francisco Sagasti, "Remarks on the Transition to a New International Scientific and Technological Order," in *Scientific and Technological Innovations, Self-Reliance and Co-operation*, Proceedings of the 8th International Conference of the Institute for International Co-operation, April 1976, University of Ottawa; and "Hacia un desarrollo científico-tecnológico endógeno para América Latina," paper presented at the SELA-CONICIT Seminar, Caracas, May 1978, subsequently published in *Comercio Exterior*, Vol. 28, No. 12, December 1978, pp. 1498–1504.

22. These principles are stated in UN Economic Commission for Latin America (ECLA), *Report of the Meeting of the Ad-hoc Working Group on Financial Machinery for Scientific and Technological Development* (Lima, Peru, 26–27 March 1979) E/CEPAL/1079, April 1979.

23. For a discussion of the concept of "endogenous S&T development" see F. Sagasti, *Technology, Planning and Self-Reliant Development* (New York: Praeger Publishers, 1979); "Towards endogenous science and technology for another development," *Development Dialogue*, No. 1, 1979, pp. 13–23 (reprinted in *Technological Forecasting and Social Change*, Vol.

16, 1980, pp. 321–330); and "The two civilizations and the process of development," *Prospects*, Vol. 10, No. 2 (1980), pp. 123–140.

24. M. V. Posner, "International Trade and Technical Change," *Oxford Economic Papers*, Vol. 31, 1961, pp. 323–341; G. C. Hufbauer, *Synthetic Materials and the Theory of International Trade* (London: Duckworth, 1965); R. Vernon, "International Investment and International Trade in the Product Cycle," *Quarterly Journal of Economics*, Vol. 80, 1966, pp. 190–227; S. Hirsch, "United States Electronics Industry in International Trade," *National Institute Economic Review*, Vol. 34, November 1965; and W. Gruber, D. Metha, and R. Vernon, "The R&D Factor in International Trade and International Investment of United States Industries," *Journal of Political Economy*, February 1967, pp. 20–37.

25. Organisation for Economic Co-operation and Development, *Gaps in Technology, Analytical Report* (Paris, 1970).

26. R. Vernon (ed.), *The Technology Factor in International Trade* (New York: Columbia University Press, 1970).

27. Harry Johnson, *Technology and Economic Interdependence* (London: St. Martins' Press, 1977).

28. Frances Stewart, *Technology and Underdevelopment* (Boulder, Colo.: Westview Press, 1977).

29. "Comparative Performance of US Technology," *Science and Public Policy*, Vol. 6, No. 3, June 1979, pp. 154–161 (extract from U.S. National Science Foundation, *Annual Report to Congress*, August 1978); National Academy of Sciences, *Technology, Trade and The US Economy* (Washington, D.C., 1978); Office of Technology Assessment, U.S. Congress, *Technology and East-West Trade* (Washington, D.C., November 1979); Regina K. Kelly, "The Impact of Technological Innovation on International Trade Patterns," Staff Economic Research, U.S. Department of Commerce (December 1977).

30. Fernando Fajnzylber, "Industrialización, bienes de capital y empleo en las economías avanzadas," *Comercio Exterior*, Vol. 30, No. 8, Agosto 1980, pp. 867–880.

31. See F. Sagasti, "Hacia un desarrollo científico-tecnológico endógeno para América Latina," *Comercio Exterior*, Vol. 28, No. 12, Diciembre de 1978, pp. 1498–1504.

32. Andean Group, *Technology and Development* (Lima, January 1979). Figures for Asia and Africa were provided in an addendum to the report prepared in March 1979.

33. Diego Pizano and Guillermo Perry (in collaboration with Francisco Sagasti), "The Scientific and Technological Dimensions of the New International Economic Order: An Exploratory Study," Draft report prepared for the International Foundation for Development Alternatives (IFDA) as part of the Third System Project, Bogota, April 1979.

APPENDIX A
Basic Features of the ITINTEC System

Introduction

The ITINTEC [Instituto de Investigaciónes Tecnológica Industrial y Normas Técnicas] system is an attempt to deal comprehensively with the problems of industrial technology policy in an underdeveloped country. It is a multiple-function organization that operates several policy instruments to develop technological capabilities in Peruvian industry and that has great flexibility with regard to the use of funds to support technological research.

The Organic Law of ITINTEC is assigned the functions of promoting, supervising, and carrying out industrial technological research; of preparing national technical norms and standards and of improving quality control in industry; and of performing additional activities such as providing technical information for industry. As a result of the reorganization of the Ministry of Industry and Tourism in late 1974, the functions of providing technical information, documentation, and extension services for industry were expanded, and the functions of dealing with industrial property and of negotiating and registering licensing agreements were transferred to ITINTEC. During the years of operation of ITINTEC the Board of Directors has also defined complementary fields for action covering areas such as engineering and industrial design, export of technology, training of personnel for technological research, and the formulation of industrial technology policies. Furthermore, ITINTEC has also developed very close working relationships with other organizations in the industrial sector and with institutions that perform similarly to the ITINTEC in other sectors. In sum, ITINTEC has become the executive agency for the formulation and implementation of industrial technology policy of Peru.

Principles for the Functioning of ITINTEC

In the first place ITINTEC will use the existing capacity for technological research in enterprises, universities, and research in-

Extract from Chapter 7 of Francisco Sagasti's *Technology, Planning and Self-Reliant Development* (New York: Praeger Publishers, 1979).

stitutions to the fullest extent, seeking to turn them into "centres for the generation of technology." Of particular importance are the technical capabilities of industry, which have lain dormant for many years, and which could be directed towards the identification of technical problems which require systematic and imaginative solutions. This implies a belief in the existence of a "hidden capacity" for technological research, which should be uncovered and used to enhance the technical level of Peruvian industry.

Second, in accordance with the National Development Plan, ITINTEC will decentralize its activities, establishing a nation-wide network of entities for the generation of technology. This implies not only spreading funds for research throughout the country, but also locating technology centres in various parts of the country to perform technical activities to support industry, both in the region where the centre is located, and at the national level.

Third, the promotion of a demand for local technology is one of the key principles that orients the operations of ITINTEC. This is done through the involvement, right from the beginning, of the users of the results in the formulation of technological research projects, and by forging links between users and producers of technological knowledge. This is of particular importance in the case of industrial enterprises which have remained relatively isolated from the development of technology research institutions until now. Also, the organization of a system, by means of which a certain proportion of an enterprise's income must be spent on research, generates an incentive for the enterprise to look at its technical problems more closely, and hence an increased demand for local technology and technological services.

Other principles which govern the activity of ITINTEC include the need to intervene directly and actively in the process of importation of technology, seeking to link this process with the production of domestic technology. Given that the great majority of the technology used by industry comes from abroad, this activity is of great importance for the definition of research projects that will enable the enterprises to adapt, modify, and absorb the technologies imported from abroad. This would lead to a progressive participation of local technological institutions in the evaluation of imported technology, the disaggregation of the technological package, and the selection of the most appropriate technologies to specific conditions in the country. Also, whenever possible, ITINTEC will seek to provide direct assistance to small and medium-scale enterprises, which lack the technical personnel to perform research tasks.

Perforce, ITINTEC will have to function as an "invisible university" for training personnel for industrial technological research. Enterprises and institutions carrying out research projects within the ITINTEC system are encouraged to include students from the universities in their projects so that they participate actively in the research project and "learn while doing." This also implies the need for acting as an agent to relate high-level specialized professionals with industries that might require their services for a specific problem.

Finally, given the magnitude of the task involved, and the dangers of growing too fast, ITINTEC will expand its activities gradually, carefully balancing short-term achievements with long-term objectives. ITINTEC will also avoid growing excessively with a centralized management structure, and it may eventually turn into a "conglomerate" which covers a variety of activities related to industrial technology policy. In that sense what is sought is a balance between comprehensiveness in the formulation and implementation of industrial technology policies, and the efficiency of a small organization with specific functions to perform.

Industrial Technology Research in ITINTEC

An underdeveloped country has to give the concept of "research" the content that fits its own situation. For the Peruvian case, Isaías Flit, the former Director General of ITINTEC, has defined industrial technological research as "the application of imagination with scientific rigour to the solution of a concrete technical problem in industry." In this way, the spectrum of technological activities covered is expanded greatly, and the usual discussions on applied versus basic research are avoided. For example, if the solution of a concrete technical problem in industry requires the performance of research tasks of a fundamental nature, they would be included within the scope of ITINTEC. This definition also puts emphasis on the source of research projects, namely, the concrete technical problems of industry, and on the two key components of research: imagination and scientific rigour.

Financial Resources for the ITINTEC System

Bearing this definition in mind, it is possible to understand better the orientation of industrial research activities of ITINTEC. Fig. 4A.1 summarizes the way in which funds can be used within the

Figure 4A.1 Uses of ITINTEC's Research Fund

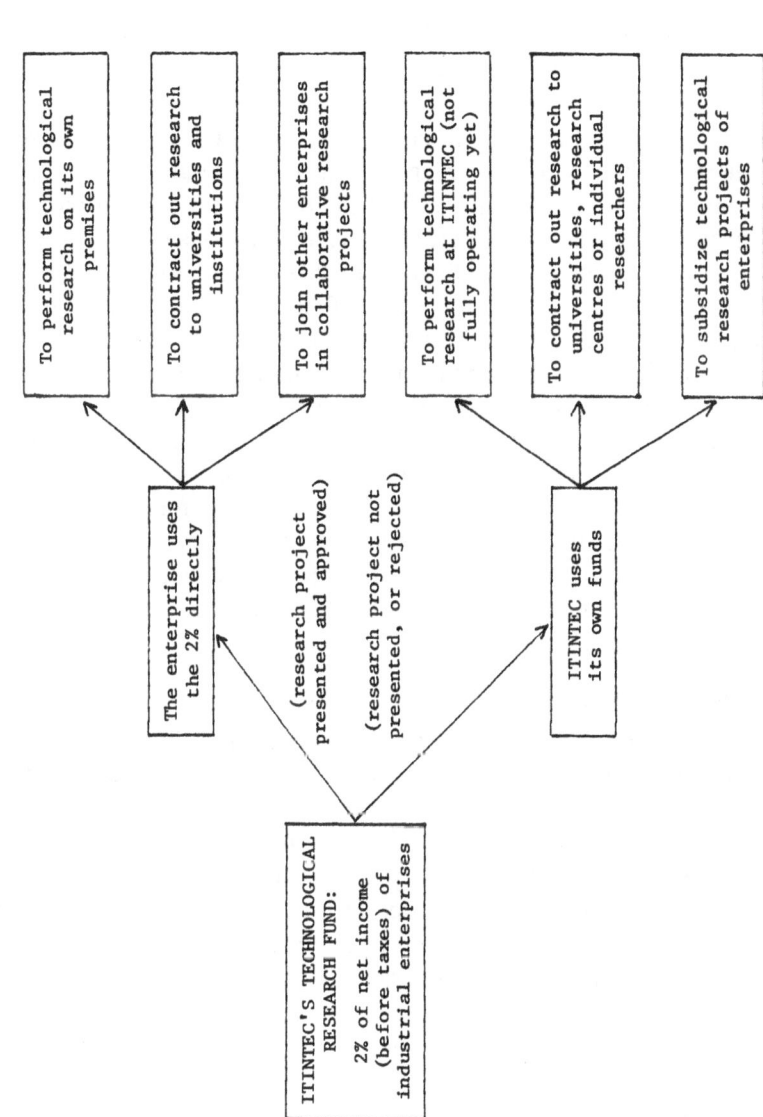

ITINTEC system of industrial technological research. Every industrial enterprise is required to put aside 2% of its net income before taxes for the performance of technological research. The enterprise has the first option to use the funds if it decides to present a research project and ITINTEC approves it. The enterprise can perform the research on its own premises if it has the manpower and equipment necessary, or it can contract out the research to some other entity primarily universities. In this case the enterprise and the research centre sign an agreement the general terms and conditions of which are specified by ITINTEC in a model contract. If the enterprise decides not to present a research project, or if the project presented is not approved, then the 2% of net income before taxes has to be deposited at the National Bank in a special account in the name of the research fund of ITINTEC.

With the funds gathered from enterprises which do not present projects, or whose projects are rejected, ITINTEC can carry out its own projects, it can contract out projects with universities and other centres, or it can subsidize research projects that are being carried out by enterprises. In this latter case, preference is given to enterprises undertaking joint research in collaboration with others, and to medium and small enterprises that have worthwhile research projects but whose funds are not sufficient.

Project Proposals

To guide enterprises in the formulation of research projects, ITINTEC has prepared a manual specifying the structure and content of the proposals to be presented by industries, universities, or any other entities. The proposals go through an evaluation process by the technical staff and, depending on the amount of funds involved, they are either approved by the Director General or by the Board of ITINTEC. In addition to requesting research projects from the enterprises, ITINTEC also asks individual researchers and research organizations to submit proposals which could be financed out of ITINTEC's fund. Once a project is approved, a contract is signed between ITINTEC and the entity that will carry it out. It specifies the terms and conditions of the agreement, establishes a detailed programme of expenditures, and defines reporting procedures. The projects are monitored by ITINTEC's technical staff, the progress of the project followed and final results evaluated upon completion.

ITINTEC's direct research activities had not been fully implemented in 1976, and there was a plan for creating several "technology centres" throughout the country, in which ITINTEC would carry out its own projects. An intermediate solution was also considered, by means of which ITINTEC personnel would do research using the equipment of other organizations or enterprises.

Basic Characteristics of the ITINTEC System

The system outlined in the preceding paragraphs has several characteristics that make it rather interesting for the purpose of promoting the development of industrial technological capabilities in less developed countries.

First, it provides a "protected market" for research and development. By specifying that funds can only be used to finance research projects (and by making ITINTEC the guarantor that funds are used for this sole purpose) it generates an effective demand for technological research. Given that enterprises face the option of using the funds for technological research, or turning them over to ITINTEC, there is an incentive to examine the technological problems of the enterprise. If the management is already convinced of the value of research this would be seen as an additional encouragement; and if management is indifferent to technological problems, this may generate a concern for improving technical capabilities.

Second, a decentralized system is established whereby the definition of research projects is spread throughout industry, avoiding excessive centralization in the definition of research priorities and projects. The presumption is that industrial enterprises know better than anybody else their own problems and are able to formulate research projects which truly respond to their own technological needs. Together with the guidelines for the preparation and presentation of projects, enterprises are further informed, through a set of criteria, as to what constitutes a worthwhile project from the point of view of ITINTEC. During the phases preceding the formal presentation, and sometimes even after this, there is a continuous dialogue between ITINTEC staff and enterprise management, in order to arrive at the definition of research projects which could be approved by ITINTEC. The main concern is to give enterprises as much assistance as possible so that they can develop their own projects or, alternatively, to develop their own capacity for identifying and defining terms of reference so that the projects can be carried out by specialized research agencies. All of this implies that the basis for the activities of ITINTEC is the research project, and that wideranging open-ended programmes, preliminary ideas, and unstructured research programmes with no clear contribution to enterprise and national objectives are not accepted. The performance of technological research projects, or the capacity for specifying terms of reference for other entities, would lead to an increase in the technology absorption capacity of industry. To allow for projects that last more than one year, the ITINTEC system provides that an enterprise can allocate its 2% for up to five consecutive years for the realization of a particular project.

Central to the operation of the ITINTEC research system is the idea that contractual arrangements for specific projects should be used, rather than the granting of subsidies or the allocating of funds to open-ended research projects. The idea is to develop the habit of "contract research" by making enterprises pay for a specific service, and therefore become concerned about the results they get. On the other hand, this would also force the research institutions to deliver the goods in accordance with the objectives, terms, and conditions specified in the research project and the contract. This is of particular importance in view of the fact that the 2% system provides a stable source of funds, free from budgetary negotiations, and which could perpetuate a situation where irrelevant research is supported indefinitely.

Another key characteristic of the ITINTEC system is that the state participates actively in the support of industrial research on a decentralized basis. In effect, the 2% is computed before taxes and therefore the state is foregoing the income it would have obtained if the 2% were computed after taxes. In some cases, for the relatively large enterprises, this could reach almost one half of the 2%. Together with the financial support, there is the right of the state to participate in the use of equipment and materials, and in the results generated by the projects. To this end, ITINTEC follows the policy of avoiding the duplication of expensive research equipment by orienting research projects of the same type to a particular centre, or by asking that equipment be put at the disposal of other enterprises or research centres which require it. ITINTEC carries an inventory of equipment purchased with the 2% fund as a means of putting into practice this policy.

The situation is more complicated with regard to the ownership of results (primarily patents), and in this respect ITINTEC has followed a very flexible policy, treating each research project on its own merits. Clearly, there may be some instances where enterprises that have obtained worthwhile results should derive some advantage over their competitors; but on the other hand, there might be cases in which the knowledge produced is of too great importance to be used by one entity alone, particularly where research projects are carried out by universities and specialized research institutions. This issue also has implications with regard to the financial returns to the results of research activities, and here again, ITINTEC follows a flexible policy, so that royalties might be shared in various proportions by ITINTEC and the entities carrying out the research.

The existence of a research fund directly at the disposal of ITINTEC provides the institution with the opportunity to reallocate funds for technological research in accordance with the needs of industrial development. In particular, this allows ITINTEC to fill the gaps where existing enterprises do not carry out technological research activities. In this respect, ITINTEC has developed, in close co-ordination with the sectoral planning office of the Ministry of Industry and Tourism, approximately 50 profiles of research projects on which proposals have been requested from universities and research centres. These projects refer to natural resources available in the country, to areas where new investments are planned, to specific problems which require urgent solutions, and to the development of research projects necessary to provide a technical infrastructure for industry as a whole.

The ITINTEC system of industrial technological research provides for widespread inputs into the process of defining priorities for industrial research. The sources of priorities and of research projects in ITINTEC are the following:

a. projects presented by enterprises, which respond to their specific technical needs;
b. projects developed jointly with the sectoral planning office of the Ministry of Industry and Tourism, or with the planning offices of other ministries, which respond to the needs of national development;
c. proposals by universities, research institutes, or individual researchers, which respond to the opportunities that they see to exploit economically a particular line of research;
d. projects that arise out of specific short-run problems and which must be solved rapidly by a contingency technology research project; these arise out of specific demands of government agencies as a result of urgent problems;
e. projects that emerge out of ITINTEC's own planning effort; these respond to anticipated technological problems in areas where ITINTEC must intervene;

   f. projects arising out of anticipated technological problems that international commitments may impose on Peruvian industry, such as the case of industrial programming in the Andean Pact;

   g. projects arising out of basic research results which show an economic potential for their application.

This scheme ensures diversification of the sources of research proposals and of research priorities. It is ITINTEC's task, and particularly that of its Board of Directors and Director General, to harmonize and consolidate projects arising out of these sources into a coherent whole, with the aims of attaining the objectives of developing technological capabilities in industry and of acquiring a capacity for autonomous decision-making in matters of technology.

APPENDIX B
Some Thoughts on the International Development Research Centre (IDRC) Experience

It seems that the few original experiences that IDRC has launched in the field of S and T have given interesting results.

Why?

A few possible answers could be listed:

IDRC has managed to maintain a restrained forward-looking attitude with regard to the research it supports in the developing world. It has been able to anticipate future problem areas which require new knowledge to be generated through research, but without imposing them on the developing countries. This curious mixture of anticipation and restraint is probably unique to IDRC. Some donor agencies prefer to define the problems they know are or will be important and support research on them, even if the developing countries are not ready or willing to work on them. Others prefer the comfort and safety of research priorities expressed by developing country governments (which are generally obsolete) and abandon any forward looking perspective. This capacity to anticipate and suggest possible research areas, in a climate of mutual respect and meaningful dialogue with scientists and policy makers from the Third World, is what has given IDRC wide acceptance.

The Centre is unique in that it maintains a global perspective in its operations. It is capable of putting researchers and policy makers from anywhere in the globe in contact with each other so as to generate a learning process, particularly among Third World countries. The Centre has managed to initiate the process of linking up researchers and policy makers from all the regions of the Third World, who identify with each other because of their common problems and interests.

However, the global perspective must respect the heterogeneity of regions and individual countries. This implies an approach that balances local and global research interests and which combines

Extracted from a paper by Louis Berlinguet, "Views on the Transfer of Technology Based on Recent Programs in the Third World," presented at the OECD Workshop on Scientific and Technological Cooperation with Developing Countries, Paris, April 10–13, 1978.

"learning while doing" with "learning from others." Of particular interest is the transfer of skills, research results and knowledge from one developing region to another, which the Centre has been actively promoting.

The IDRC is prepared to give a chance to the developing country researcher to "do it on his own," to prove himself, while at the same time safeguarding against possible excesses and abuses. This implies a belief in the existence of a "hidden research capacity," which has not been given the chance to develop, in the ability to learn rapidly while doing research, and in the willingness to learn from the experience of others in the Third World.

One important issue in this respect is the need to adjust the types of support given to the characteristics of the region and country, supporting research institution versus supporting individual researchers.

The Centre has a hybrid legal nature which makes the projects it supports neither governmental nor private; it can be considered as an international organization, but it is not multilateral. In principle, it appears to have the best of all worlds, although this has created some trouble in countries which find it difficult to place the Centre. The participation of Board members from developing countries strengthens this hybrid position and gives more credibility to the Centre. In short, the institutional status of IDRC and the type of financial support given by the Centre allows researchers and policy makers to talk freely, leaving aside their official government positions, but within the framework of relevance to real development problems.

The Centre has managed to maintain a flexible set of operating and administrative procedures, adjusting to the needs of each country that participates in a project. There are no immutable rules on how a project should be managed and a large degree of freedom is given to the researchers in the developing country. This is closely related to the nature of the "monitoring" process through which the Centre follows the progress of each project.

The IDRC supports problem-oriented research. It constantly asks "What difference does it make?" whether a problem is addressed and whether or not the solution is liable to have a significant impact on the development of LDCs. Heavy emphasis is placed on adaptive and applied research, with basic research being undertaken when it is clearly necessary to overcome some identified obstacle — not as an end in itself.

Proposals Submitted by the Group of 77 to the United Nations Conference on Science and Technology for Development

5. Global Financial Arrangements

It is necessary to have an effective financial mechanism or system to assist in the implementation of the measures recommended in this action programme, including regional and subregional projects. This mechanism or system should have the following characteristics:

a. It should be used for the strengthening of scientific and technological capacities of developing countries, including the acquisition of technology;

b. It should mobilize and channel all types of financial resources, particularly from developed countries;

c. Its resources should be substantial and supplementary to the resources that now exist, and furnished on a predictable, automatic and continuous basis;

d. The volume of its financial resources should be sufficient to contribute effectively to the implementation of the measures contemplated in the Programme of Action. An initial target must be determined;

e. It should be set up within the United Nations system, in such a way as to be duly co-ordinated with the competent organs of that system, and independent in its operation;

f. The developing countries should be enabled to participate fully in its operation.

6. Global Financial Arrangements

C.20. The Conference decides to recommend to the General Assembly of the United Nations that it establish, at its thirty-fourth session, a financial system which should become operational in January 1981 and should be known as the United Nations Financing System for Science and Technology for Development, on the basis of the following:

A. Objectives

C.21. The system shall finance a broad range of activities aimed at strengthening the endogenous scientific and technological capacities of developing countries and in particular to assist in the implementation of the measures envisaged in this Programme of Action. Those activities shall be complementary to and supportive of the national efforts of the developing countries in the fields of science and technology. It shall be a vehicle for the mobilization, co-ordination, channelling and disbursement of various types of financial resources.

B. Resources of the System

C.22. The Conference agrees that in determining the nature and level of the resources of the System the following considerations should be taken into account:

a. The asymmetry of the technological capacity between developed and developing countries;

b. The need for predictability and continuous flow of financial resources;

c. The need for substantial resources in addition to those that now exist within the United Nations system;

d. The need for untied external resources for the scientific and technological development of the developing countries.

C.23. In view of the above considerations, the Conference decides as follows:

Level of the total financial resources of the System

a. Bearing in mind the necessity to start the operations of the System in January 1981 with sufficient resources, the annual target for the financial resources of the System should attain progressively at least $US 2 billion by 1985, with the aim of reaching at least $US 4 billion by 1990.

Nature of the resources of the System

b. The resources shall be established on the basis of assessed contributions as well as voluntary contributions;

(i) Contributions from developed countries calculated on the basis of a percentage (to be determined) of the average quinquennial surpluses of developed countries in their balance of trade in manufactured goods with the developing countries. The quinquennium to be considered for this purpose is the one ending three years before the date when the contributions become due. The volume of resources accruing from these contributions shall constitute a proportion substantially superior to 50 per cent of the total resources of the System, in accordance with the target set in C.23a. above;

(ii) Contributions from developing countries commensurate with the level of their financial, scientific and technological capacities;

(iii) Other contributions from developed countries, foundations, intergovernmental and non-governmental organizations and institutions.

Allocation of resources of the System

c. The resources of the System should be allocated to the various activities identified in this Programme of Action, including regional, subregional and interregional activities. The Intergovernmental Committee will establish guidelines for the allocation and the distribution of the resources for the System, within the framework of the priorities of the developing countries, for implementing in particular different types of projects and programmes of direct relevance to the developing countries, taking into account *inter alia* the need to strengthen the collect self-reliance of the developing countries, the need to take special measures to meet the urgent and specific problems of the least developed, land-locked, island and most seriously affected developing countries as well as other criteria to be adopted by the Intergovernmental Committee. Additional criteria for the allocation of resources should provide *inter alia* for a part of the resources of the System to be applied to high-risk research and development science and technology projects at the national, regional, subregional and interregional level and to provide support for the developing countries in obtaining financial resources from other sources.

C. Other Financial Resources

C.24. The System may enter into arrangements with international, regional and other public and private financial institutions with a view to the generation and channelling of additional resources to the developing countries for scientific and technological activities including research and development, and the commercialization and acquisition of technology.

C.25. The resources to be derived from these arrangements should be supplementary to the System's own resources. Such resources could be provided by:

a. International and regional financial institutions;

b. Public and private banks of national, regional and international types;

c. Public and private corporations;

d. Other public and private financial institutions.

C.26. Additionally, the System may use other resources, such as;

a. Resources accruing from concrete progress towards general and complete disarmament, including the urgent implementation of the already agreed disarmament measures;

b. Resources accruing from the proposed "international labour compensatory facility" related to the reverse transfer of technology.

D. Institutional Arrangements for the System

C.27. The governing body of the financing system shall be the Intergovernmental Committee on Science and Technology for Development which should make recommendations to the General Assembly at its thirty-fifth session on the managerial, functional, operational and administrative modalities of the financing system.

The Conference recommends that at its thirty-fourth session, the General Assembly of the United Nations request the Director-General for Development and International Economic Co-operation, in co-operation with the executive heads of the competent United Nations organs and specialized agencies, to undertake a comprehensive and action-oriented technical study, on the basis of the principles enunciated above, for the implementation of the activities of the System. The study should be addressed in particular to the following aspects:

a. Detailed quantification of assessed and other potential sources of funding;

b. Draft statutes of the System and other organizational arrangements, including association arrangements with other institutions and organizations;

c. Specific steps for carrying out the activities of the System.

The study should be submitted through the Intergovernmental Committee on Science and Technology for Development to the General Assembly at its thirty-fifth session for its decision.

5
The Role of Women in Science, Technology, and Development: Perspectives on UNCSTD

Pamela D'Onofrio-Flores

1. Introduction: A Comprehensive Approach to the Study of Women in Development

Women suffer from many kinds of exploitation, as unpaid labour in households[1] and as paid labour in jobs that have been earmarked as female occupations and are located in the less dynamic sectors of the economy. These sectors are peripheral to central areas of economic investment and productivity and are the least penetrated by advanced technology. As workers often engaged in this marginal employment, women are typically paid less than men. Thus there exists a division of labour by sex that is detrimental to women because it assigns them nonpaid work in domestic-related activities or low-paid work in marginal sectors of the economy. The subordinate position of women in the economic division of labour contributes to the attitude that women are inferior members of the work force. As a result, women in many cases are paid less than men even when they perform the same work. These forms of exploitation are made worse when women are also members of racial minorities or citizens of semicolonial nations. When this is the case, women are subject to a tremendous amount of economic, legal, and social disability.

No analysis of the situation of women is accurate without explicit reference to the wider social context that includes social classes.[2] Many studies, including those by Arizpe, Chinchilla, Kandiyoti, and Stoler,[3] show that both women and men of the lower socioeconomic classes face formidable obstacles to equal educational, economic, and political opportunities. These studies also show that the discrimination faced by members of the lower socioeconomic classes and the discrimination faced by women are related in that the capitalist form of economic organization both widens the sexual division of labour and increases class inequalities. Thus the major burden is placed on women of the working classes, who are exploited as members of a subordinated class and sex. These studies conclude that the division of labour by sex must be examined in its relation to class in the advanced capitalist countries as well as in the developing ones.

However, the struggle of women is more than a class struggle, because even within this class differentiation women, as women, play a subservient role. In this regard the assumption that the replacement of one class by another in the leadership of the state machinery will reverse male dominance is misleading. Although this is a necessary condition for the advancement of women it is not a sufficient one. The structure through which services are provided, the structure through which goods and services are consumed, and the ideological superstructure that perpetuates inequalities of all kinds, including male dominance, must also be transformed.

The phenomenon of technological discrimination against women cannot be viewed apart from these structural questions, nor can it be seen as separate from modes of production, land tenure systems, ruling ideologies, and legal and cultural biases.[4] Consequently, understanding women as social beings demands the exploration of the relationships that exist between community and workplace, public and private life, family and economy, and reproduction and production.[5]

To discuss the impact of scientific-technological change on the role of women in development means locating women in time and space. Women are part of a world system of uneven economic, political, and social relations, and women's role in society is determined, in large part, by these uneven relations. Moreover, the exploitation of women is closely related to the wider oppression and marginalization of entire racial and ethnic groups and social classes. Consequently, any attempt to examine the issue of women's exploitation must be linked to an understanding of how their exploitation is related to other forms of marginalization and oppression of societies and of other groups therein.

The attainment of sexual equality requires identification of the mechanisms that have the capacity to fuse the social movements of women, as women, with the drives of women and men against oppression as workers and members of exploited races, social classes, and nations. This suggests an approach that can cultivate understanding of the existing relationships between inequalities at the sexual, national, regional, and international levels, and the ways in which they are mutually reinforcing.

A strategy focus that considers women as a human resource will not modify the relations of dominance upon which their participation is based. The incorporation of women into nationally defined productive activity will not necessarily imply a change in their subordinate role. Contemporary forms of exploitation of the female work force might only be exaggerated. The type of development involved and the method of female participation in it must be specified.

Eliminating current sexual inequalities warrants structural changes in the social relations both in the economic sphere and in the closely related sphere of power. The springboard to an understanding of the situation of women when science and technology advance and to subsequent positive change lies in the comprehension of the historical roots of women's exploitation as well as in the understanding of its contemporary manifestations. Knowledge of the sources of female oppression is a critical prerequisite to devising strategies and tactics aimed at eliminating that oppression. As these targets for change become revealed so will the strategic focus sharpen. Similarly, as the processes through which women are exploited surface, so will the tactics necessary to counter them.

A focus on women's historic position vis-à-vis the productive, political, and social structures is critical also because women's participation in national development and in scientific-technological development requires not only opportunities to contribute but also opportunities to overcome the burdens of marginalization. Out of the historic marginalization of women from decision making in the social, economic, political, and technological spheres there have come different and often constructive critical perspectives, which can be cultivated as the bases for creative alternative structures and solutions to global problems. However, attempts to clarify the history and sources of women's exploitation collide with a barrage of myths concerning their roles in history, economic relations, and psychology and their relation to socioeconomic, political, and scientific-technological change.

2. The Demystification of Women and Development

The division of labour by sex is most widely attributed to "tradition." Inadequate attention is paid to the historic evolution in occupational structure and changes in the allocation of work by sex. Stereotypes regarding women's subordinate "traditional" role and status originate from the same unilinear view that considers all cultural traditions of the developing countries as antithetical to development.

In recent years there has been considerable controversy among social and economic historians, anthropologists, sociologists, political scientists, and other specialists concerning the impact of modernization on the role of women in society. An increasing number of studies have come to challenge the conventional wisdom that women have been traditionally exploited in developing societies and that "modernization" is the key to the reversal of that trend.

For example, the prominent historic role of women in trade has been much remarked on in certain societies, including many of those of West Africa, the non-Hispanic Caribbean, and parts of mainland Latin America.[6] Sidney Mintz stressed the point that with modernization there occurs a separation of the internal market system from the channels by which agricultural commodities reach export houses. This is not a difference between subsistence crops and cash crops, but rather one between locally produced goods for local consumption and those produced for export. Such a division in many cases circumvents female intermediaries even in societies where women have been traditionally important and active in trade, such as Nigeria and Haiti. Mintz concluded that internal market systems, where female activity is concentrated, may expand in response to economic growth of other kinds derived from modernization but that these wider, more complex kinds of economic growth will probably provide women with fewer opportunities than those available within an expanding internal market system.[7]

Contemporary attention has shifted to the examination of the structure of women's exploitation and the effects of colonialism, neocolonialism, imperialism, and capitalist- and socialist-oriented development strategies on women's economic, political, and social role and status. Accumulating evidence supports the idea that female status was relatively free of exploitation in many precolonial societies, that women's function as principal producers gave them a dominant social role in agrarian-based societies; and that the ideology and structure of male dominance were frequently introduced as corollaries of colonialism, rooted in the capitalist form of economic and social organization.

According to the anthropologist Eleanor Leacock:

> The image of women as naturally the servitors of men and men as naturally the dominators of women reinforces the myth that traditional family relations in Third World nations were based on male dominance that characterized Europe, where the calvinist entrepreneurial family was of great importance to the rise of capitalism. The idea of women's autonomy is then presented as a Western ideal foreign to the cultural heritages of Third World peoples.
>
> The fact is however that women retained great autonomy in much of the pre-colonial world and related to each other and to men through public as well as private procedures as they carried out their economic and social responsibilities and protected their rights. Female and male sodalities of various kinds operated reciprocally within larger kin and community contexts before the principle of male dominance within

the individual families was taught by missionaries, defined by legal statutes, and solidified by economic relations of colonialism.[8]

This can be seen most clearly in Africa. Rousseau-Mukenge, in her conceptualization of the role of African women in development observes that

> The need to expropriate the wealth of Africa had its economic sources in the development of European capitalism. Part of the way in which human resources were mobilized under this system was culturally rooted in the European capitalist definition of women. Simply put, women in the European economic and cultural context were seen as the procreators and nurturors of labour for capitalist production, and also as a temporary reserve labour supply. Women in the colonial capitalist economic enterprise were seen no differently. Hence, African women were regarded and treated as European women in their relationships to the evolving dependent social, economic, and political structures in Africa.[9]

Further evidence suggests that although contemporary development may afford political and professional roles to some token women, it continues to undermine the status and the autonomy of the majority of women.[10] There is no inherent positive correlation between modernization, technology, and the improvement of women's economic, political, and social status. The consequences of the massive introduction of foreign technology into the economy vary depending upon its social utilization.

In this context, the mode of production and the development strategy undertaken are of considerable importance in appraising the effects of scientific-technological change on the role of women in the development process. For example, Muller and Fleischer asserted that the establishment of agricultural production cooperatives improves the status of women in agriculture by specifying that women farm members are owners enjoying the same rights as male farmers. "In the German Democratic Republic, women in agriculture account for 13.1 per cent of all executives and 44 female members of cooperative farms are members of Parliament. About 15 per cent of all women working in agriculture have a license for driving trucks and tractors and a permit for operating large machines and milking units."[11]

China serves as an example of the close link between the development strategy adopted and the socioeconomic roles of women:

The Chinese Communist Party has argued that the presence of female solidarity groups which are devoted to protecting and furthering women's economic and political interests were necessary to draw women into new economic and political activities and to facilitate their access to and control over, not only the products of their labour, but also the economic and ideological resources of society. . . . During the successive government policies to displace individualized peasant production with collectivized agriculture through land reform, the collectivisation and communisation of agriculture and the establishment of rural industries and projects of capital construction in the countryside, it was the women's groups in the villages which encouraged women to take advantage of the new opportunities available to them to take a full and wide-ranging role in production. Small cooperative production units of rural handicrafts and all-women production teams working in the fields have often coincided with the basic organisation of the local female solidarity groups in the villages. In the production units, it was often these groups which made arrangements for and encouraged women to learn new skills and break into new spheres of work which were traditionally male preserves. . . . The female solidarity groups also acted as pressure groups to expedite the implementation of government policies to accommodate women's biological role, introduce public health measures to reduce infant mortality and the means to give women control of reproduction and to end discrimination against women in favour of the policy of equal pay for equal work.[12]

3. Some Sources of the Marginalization of Women from Productive Activity and from Modernization

The failure of the development process to benefit women is a function of unequal exchange resulting from disparities in labour productivity and economic, institutional, and political power. Much attention has been given to the way unequal exchange operates between rich and poor countries, between urban and rural areas, and between social classes. Attention must now be turned to the similar way in which the same process of unequal exchange operates between the sexes.[13]

By way of illustration, monetization and market incorporation, two facets of "modernization," give rise to new institutions. Markets require a place containing appropriate supportive facilities, and productivity changes involve the channeling of credit and technical information. Whether cooperatives or marketing agencies, these new institutions have, in almost all cases, given prominence to the male head of the household as the selling point in domestic units.[14] "Modernization" thus creates new exchange relationships resulting from the unequal distribution of institutional and economic power between the sexes, operating to the detriment of women. Additionally, given the nature of women's work in the home and in self-provisioning food farming as unmonetized, relatively uncapitalized, and low-productivity activity, the inequality of exchange between the relative labour of the sexes will undoubtably increase.[15]

Modes of appropriation are a function of exchange relationships, which are influenced by productive relations. Both sets of relations are profoundly influenced by oligopolistic tendencies in the supply of land, capital, and in marketing relationships.[16] Therefore, there is nothing to prevent women from becoming worse off as "modernization" progresses.

According to Saffioti, contemporary discrimination against certain groups in society, including women, is not due to the persistence of unequal traditional legal and ideological structures prior to capitalism. In her view, discriminating and oppressive structures are an inherent part of capitalism and are supported and nurtured by the economic infrastructure of capitalist society. "Thus at the level of mode of production the situation of women is constant: relegation to an inferior condition on the ideological plane to justify their fundamentally economic marginalization."[17]

In the capitalist form of organizing economic production, the end of production is the exchange for profits, and concern for domestic consumption needs is subordinated to short-term gains for an enterprise or an industry. In addition, women's biological role in reproduction is linked to the social responsibility of nurturing, socializing, and maintaining present and future generations. When this conceptualization of women's role is combined with a development strategy that gives greater importance to growth in production than to social development, the employment of women in the advanced sectors of the economy is "uneconomic," as it entails a heavier investment in social services (e.g., child day care facilities).

The additional labour expended by women within the household that is external to national definitions of productive activity is economic and essential to capitalist development. If all of the goods and services required for the maintenance and reproduction of the labour force that are provided by women were to be purchased as commodities, the magnitude of the wage would have to be increased or the standard of living of the working class would be reduced. The domestic unpaid labour of women in this way allows the payment of wages below the actual costs of production and reproduction of labour power. Thus the capitalist economy increases its savings through the productive efforts of women in the household.[18] These savings, derived both from unpaid household production and women's work in low-paid jobs in the marginal sectors of the economy, can be invested into the central sectors of the economy.

The household also maintains women as a reserve labour force available to join nationally defined production when the situation demands their inclusion. This eases the social tensions of rising unemployment, enabling capitalism to survive its chronic cycles of inflation and depression. During a decline in the demand situation wages are depressed to a starvation level, or in the opposite instance, when the demand for labour is high, employers draw on the reserve supply of underemployed women who have been conditioned to withdraw to the home during recessionary periods. For example, historically war has brought North American women their greatest economic advances. During every period of national emergency, women have been utilized in differing productive capacities, but thereafter have been subject to treatment as a marginal segment of the work force.[19]

Women in the work force in industrially developed countries are experiencing the same insecurity as the first to be fired and the last to be hired that women in the developing countries face, one of the differences being the extreme nature of unemployment and the narrow margins of welfare benefits in the developing countries. In the developing countries, women's economic functionality centers primarily on the maintenance of a low-value labour capacity through their production of subsistence foodstuffs.[20]

The global variation in the equal participation of women in the labour force depends not as much on the question of industrially developed as compared with developing but on the commitment to a full employment policy as compared with the treatment of labour as a commodity equivalent to other factors of production and subject to the same fluxuations in demand.[21]

The rate of female activity in the West has shown increases in the most industrialized countries, particularly in Denmark, Sweden, and Finland, with 58 percent, 59 percent, and 63 percent respectively, and in the United States of America. The large increases in demand for women workers in the United States since World War II have been in large part a joint product of the sex-typing of jobs and the fact that the particular jobs earmarked as female occupations are located in the expanding service sector. This is also true of countries like Sweden and the Netherlands, which have already become full-fledged service economies with more than half of their labour force employed in the service sector.[22]This suggests that the difference in wages between the sexes is in large part due to the concentration of women in low-paying occupations as well as to the secondary role of work in their lives.[23]

There is a difference in the female economic activity rates for developing and industrially developed countries; there is an even greater differential between the Soviet Union (51 percent) and the United States (34 percent). The female economic activity rates in the labour force in Poland, Romania, the German Democratic Republic, Bulgaria, the USSR, and Hungary are an average of 10 percent higher than those of the market economies. Nor is the participation of women as restricted to female-segregated jobs, such as secretarial work, nursing, and precollegiate education as in most of the market economies.[24] However, although the sex-typing of jobs and the entry of token women in high-level posts is more marked in advanced Western industrialized countries, it prevails in the East European countries as well, where some occupational barriers have been broken, but equal entry into the top levels for women in social, economic, political, or technological spheres is still not a reality. The change in the status of an occupation as a function of female entry into the profession in great numbers remains a subject for further inquiry.

As capitalist activities increase in the developing countries, primarily through multinational corporations, so the marginalization of women in relation to economic activities and to decision-making positions is accentuated. Saffioti attributed this intensified marginalization of women to the penetration of capitalism on the basis of the use of highly labour-saving technologies.[25]

4. Conventional Forms of Technology and the Exploitation of Women

Technological development and use, in their contemporary form, are components of the infrastructure for the perpetuation of assymmetrical relationships between industrially developed and developing countries, between regions, between classes, and between social groupings within classes. Vyasulu argued that technology in its current modern form is a "tool of oppression" used by an international elite class best represented by transnational corporations, which indirectly control technology and research.[26] The development and use of technology are agents of marginalization of nations and their peoples to the extent that such development facilitates the concentration of knowledge in the hands of dominant countries and social classes.

In the contexts of most countries, the choice of technologies and their implantation sharpen the already marked disparities in earnings and sociopolitical efficacy between men and women. Studies prepared for the seminar session on "The Changing Roles of Women in Rural Societies" of the Fourth World Congress for Rural Sociology (1976) all converge around the central conclusion that the processes of "modernization," i.e., industrialization, agricultural mechanization, and technological advance have been directed almost exclusively to male peasants in developing countries.[27] Moreover, there is substantial evidence to support the claim that women suffer a loss of economic authority and security as well as general status during the processes of rural modernization.[28] This is the case largely because present methods of choosing and introducing technological improvements have the dual effect of increasing the female work burden and simultaneously undermining women's rights of appropriation over the returns on their own labour.

The Mwea irrigated rice scheme in Kenya provides a documented case study illustrating how the introduction of commerical crops and the commercialization of food staples can reduce women's ability to secure an equitable share of family produce and cash income while increasing the length of their working day, ensuring their continued employment in labour-intensive, low-productivity work, and reducing their control over the family's purchasing power.

The scheme was intended to raise family income via the intensive cultivation of rice as a cash food crop. Families were transported to the irrigated settlement village and the agro-inputs were supplied to the male head of household. As a result, the domestic provision of food continued to lie within the sphere of female responsibility, but the earnings from the household's cash cropping were placed under the control of their husbands by virtue of the latters' special relationship with the settlement authorities. In the new village the plots for growing traditional food crops were smaller and inadequate for feeding the household. Moreover, the women found that they were allocating more time to the cultivation of the rice crop. The wives received from their husbands some of the crop in return for their labour but because the husbands refused to eat rice it was sold in order to buy traditional food. The wives were continually in need of cash to compensate for the inadequate traditional food plots and the reduced time allocated to their cultivation. This form of agricultural modernization detrimentally affected women in two ways: They had to work longer hours, and they lost their ability to draw upon their own resources to supply the family's food.[29]

The Bouaké region of the Ivory Coast stands as further evidence of the positive correlation between the advent of current forms of agricultural modernization and women's loss of control over the returns to family and to their own labour (in cash or produce). In the Bouaké region only 10 to 35 percent of the family income (in modernized villages) is allocated to women, as against 50 percent allocated to women in traditional villages.[30]

The proliferation of conventional forms of science and technology development and use and the concomitant acceleration of social polarities make it a matter of some urgency to understand the consequences of such proliferation for national development and what measures can be taken at the national and international levels to prevent it. In the absence of new forms of constructive social, economic, and technological relations, the New International Economic Order only offers the opportunity for more women and poor people to be marginalized labour in the modern sectors of the free world economy or to continue as rural landless labour.

5. United Nations Statements
on Women and Development

ECOSOC Resolution E/1978/34 on Women in Development and International Conferences calls upon ECOSOC to urge all governments to ensure that the topic of women and development be included within the substantive discussions of the Conferences and where appropriate be considered as a separate Agenda Item. General Assembly Resolutions 3342 (XXIX) of 17 December 1974 and 3524 (XXX) of 15 December 1975 on the integration of women in the development process urge governments to give sustained attention to the integration of women in the planning, formulation, design, and implementation of development projects and programmes. The Governing Council of the United Nations Development Programme at its nineteenth session requested that the integration of women in development should be a continuing consideration in the formulation, design, and implementation of the projects and programmes of the United Nations Development Programme. Studies and activities of the Economic Commission for Africa (ECA) Training and

Research Center for Women, the United Nations Children's Fund (UNICEF), UNDP, FAO, UNESCO, and the World Bank concerning the development, promotion, and improvement of village-level, low-cost technologies available to women seek to promote and facilitate the integration of women into modern national economic life.[31]

Findings and recommendations of the other global conferences of the decade of the 1970s also give weight to the inclusion of women in development as an issue area in the intergovernmental deliberations of UNCSTD. In its Resolution XII (Population and the Status of Women) the 1974 United Nations Population Conference requested that the United Nations organs and specialized agencies give special consideration to the impact of development efforts and programmes on the improvement of the status of women, especially in connexion with the review and appraisal of the International Development Strategy for the Second United Nations Development Decade. In its Resolutions II (Priorities for Agricultural Development), V (Policies and Programmes to Improve Nutrition), and VIII (Women and Food), the United Nations World Food Conference, also held in 1974, urged priority consideration of women in every stage of the design, planning, implementation, and evaluation of development programmes and projects.

Recommendations of the Report of the World Conference of the International Women's Year, 1975, highlighted the development of integrated or special training programmes for girls and women in rural areas to enable them to participate fully and productively in social and economic development and to take advantage of technological advances in order to reduce the drudgery of their work. Programmes in modern methods of agriculture and in the use of equipment were suggested. Recognizing women's labour in the production, processing, and vending of food, the improvement of techniques and equipment for food production, processing, preser-, vation, and conservation, as well as the distribution of these improved techniques and equipment to women in rural areas was advocated. Findings of the World Conference of the International Women's Year showed that although industrialization constituted one of the main mechanisms for female integration in development, women workers are disadvantaged in many respects because the technological structure of production in general is oriented towards men and their requirements. Its recommendation calls for examination of the situation of the female worker in food-related industry and in services.

Moreover, the International Labour Organisation, in its 1976 report *Employment Growth and Basic Needs*, stated:

> Especially in rural areas most women in developing countries are over-worked rather than underemployed and a more appropriate technology for the tasks they perform implies labor saving in order to improve the quality of their employment rather than employment creation. Much emphasis is often placed on the relief of drudgery in agricultural work by mechanization but this unfortunately in many cases reduces employment opportunities at the same time. *There is much scope for relieving the drudgery of women's household work* by the provision of accessible water points, rural electrification and *simple technological improvements in the processing and preparation of food* at home.[32]

At the recent United Nations Conference on Technical Coopera-tion Among Developing Countries in Buenos Aires, the delegate from Mozambique observed that

> We have nearly forgotten to analyze the place of women in the development process. But it is a fact that the hard core of the develop-ment problem is women. The subject needs to be approached honestly and scientifically, being careful in the use of statistics for "economically active" workers, being profoundly analytical when talking "Appropriate Technology," being aware, that it is generally women's muscle power at stake.[33]

6. UNCSTD: A Step Toward the Improvement of Female Status and Participation in Science and Technology

The United Nations Conference on Science and Technology for Development (UNCSTD), held in Vienna, Austria, in August 1979, sensitized the international community to the exisiting inequalities facing women in scientific-technological development and provided the necessary direction to include women in scientific-techno-logical decision making and implementation.

The working group on Science and Technology and the Future presented a report that included the following paragraphs adopted by the Conference:

> 21. Technological development often affects men and women dif-ferently and the introduction of new technologies has tended to have an adverse impact on the latter, thereby lessening their earnings and social status. It is therefore of the utmost interest to society that in future the full participation of women be ensured in the planning and

setting of priorities for research and development as well as in activities relating to the design, choice and application of science and technology for development. They should also be provided with equal access to scientific and technological training and professional career opportunities. In developing countries an adequate share of resources available for research and training should be allocated to the advancement of skills of women in the fields traditionally occupied by them as well as new fields.

22. Rapid development of science and technology throughout the world will depend in part on the younger men and women who can be brought into the fields and involved in decision-making bodies and given full opportunity to use their intelligence and skills. In the biosciences, for example, three steps are essential to accomplish this: (a) improved education in the ideas and methods of modern biology including the necessary grounding in physics, mathematics and chemistry; (b) creation of well equipped research laboratories in many developing countries; and (c) a much greater exchange among young biological scientists and technologists of developed and developing countries. This approach should be equally applicable to all other fields.[34]

The Conference adopted a resolution entitled "Women, Science and Technology." The resolution invites Member States to facilitate an equal distribution of the benefits of scientific and technological development, participation of women in the decision-making process, and equal access to training and the respective professional careers. The resolution also recommends that all organs, organizations, and other bodies of the United Nations system should review the impact of their programmes on women and promote the full participation of women in the planning and implementation of programmes. In addition, the resolution invites the proposed Intergovernmental Committee on Science and Technology to give due regard to the perspectives and interests of women in its activities and to review the progress in implementing the resolution in its annual reports.[35]

The Conference also adopted the Vienna Programme of Action for Science and Technology for Development. The fifth paragraph of the preamble states:

5. The ultimate goal of science and technology is to serve national development and to improve the well-being of humanity as a whole. Men and women in all groups of society can contribute positively to enhance the impact of science and technology on the development process. However, modern technological developments do not

automatically benefit all groups of society equally. Such developments, depending on the given economic, social and cultural context in which they take place, are often seen to affect various groups in society differently. They may have a negative impact on the conditions of women and their bases for economic, social and cultural contributions to the development process. This is seen to happen in industrialized as well as in developing countries. Therefore, steps should be taken to ensure that all members of society be given real and equal access to, and influence upon the choice of technology.[36]

Also, in section III.D of the Programme of Action, under the heading "Development of human resources," paragraph 99 states that "the organs, organizations and bodies of the United Nations should . . . (g) strengthen support for national efforts to promote the full participation of women in the mobilization of all groups for the application of science and technology for development."[37]

National and regional analyses of socioeconomic problems relevant to science and technology were made during the preparatory process of the Conference. The issues raised in these analyses formed the basis for UNCSTD programming. In addition to the aforementioned sections of the Vienna Programme of Action, some of these national and regional analyses contribute to greater understanding of how science and technology can have negative impacts on women and how women are discriminated against in the choice of technology, the development of science and technology policy, the direction of scientific-technological research, scientific-technological educational opportunities, and professional careers in the fields of science and technology.

For example, the Regional Paper prepared by the UN Economic and Social Commission for Asia and the Pacific (ESCAP) and submitted to the Third Preparatory Committee Meeting for UNCSTD states:

It appears characteristic of the socio-economic set-up in most of the developing countries that women rank inferior in status and opportunity to men . . . [e.g.] literacy ratios and economically active percentages are much lower in the case of women. . . . In rural areas, women work practically as bonded labour for manual operations in the fields and household chores. They work longer hours, with primitive implements and facilities. Technology has not been applied to eliminate the drudgery of their work; . . . few women can avail themselves of scientific and technological opportunities, particularly for senior positions at the executive and managerial level; only in cases of shortage of manpower are women drawn into other professions. Even in these spheres, e.g. agriculture and industry, as jobs become technology-oriented women are gradually displaced. . . . Apart from its social undesirability, the discrimination against women impedes the application of science and technology by keeping a sizable section of the population from participation in socio-economic development. The situation should be remedied through social reform and legislative action.[38]

The development of scientific and technical consciousness and participation requires a brand of education that builds an enquiring outlook, fosters analytic thinking, and provides a framework for choice and innovation. Deficiencies in female educational opportunities, qualitative as well as quantitative, particularly the exclusion of women from scientific-technological training, perpetuate inequalities between men and women in science and technology research and development manpower. The alarming extent of these imbalances is given concrete expression in many National Papers. For example, the National Paper of the Democratic Republic of the Sudan reported on a recent survey of total personnel working in scientific units in government, semigovernment, and private sectors that illustrates both the unexploited reserve of female science and technology potential and the discrimination against women scientists. Its findings were the following:

Sexwise, males outnumber females. . . . The majority of female scientists and engineers was confined to the general service and higher education sectors accounting for 88.4% of the total, almost equally divided between them. Out of the remaining 11.6% the integrated productive sector and the non-integrated productive sector accounted for 5.8% each. *The highest percentage of female scientists and engineers in relation to males in each sector* of performance was the higher education sector (9.1%), followed by the general service sector, (7.5%), the non-integrated productive sector (4.4%) and the integrated productive sector (2.6%).[39]

With regard to women and education the African regional paper observes:

One of the greatest challenges in the expansion and democratization of education — especially scientific and technological education — is the fact that in Africa the scientific and technological potential of women is rarely tapped, and when it is, it is often not used effectively. This is a worldwide problem, of course, but it attains its worst proportions in Africa. . . . This imbalance ought to be an issue high not only on the agenda of women's liberation movements but also on the list of priorities in national attempts at educational reform.[40]

New technologies often bring with them new forms of social organization that may have negative consequences for the role of women in the family and in society. It is to these unintended consequences of technology development and use that the Norwegians addressed themselves in their National Paper:

> As men increasingly are commuting between home and job, child care involves mainly the mother while the fathers' possibilities for contact with the children and for assuming tasks in the home are few. Women are thus even more hampered in their own development and frequently prevented from seeking economic independence through work outside the home. They are furthermore in a poorer position than men when it comes to participation in community life.[41]

The Norwegian National Paper also calls for the activation of "hidden" knowledge and technologies generated in local communities and based on the historical experiences and traditions of small farmers, in particular the experiences of women.[42]

The National Paper of Sweden advocates that research and development efforts at a national level should focus on diminishing the inequalities in society, activating all social strata, broadening research positions, and ensuring that women have an equal place in all the above contexts.[43]

7. Conclusion

Women have historically been and are presently integrated in economic development worldwide. However, because there is the continued perception of a dichotomy between the modern and the traditional sectors, between economic activities done for money and those done as a citizen's duty, between productive work and welfare activities, statistics still tend only to reflect activities in the modern monetary economy. As a result, women in most countries of the world are the "invisible" economic base of society and as such are excluded from the mainstream of decision making and planning, which means that they can be controlled more effectively.

The advancement of women is central to the objective of creating national and international environments conducive to constructive science and technology in the public interest. This is the case because women comprise a major segment of the urban and rural poor and because, particularly in the rural areas, women are the focal points of family welfare, not only producing most of the basics to sustain life but also transforming them and delivering them to their point of final consumption.

Now and in the future the international community should focus its attention on the following issues if sexual inequalities are to be redressed:

- The relationships that exist between production and reproduction in light of the conflict of interest between both activities;

- The interrelated dynamics of modernization, power distribution, scientific-technological progress, and the socioeconomic and political participation of women;

- Reevaluation of the present pattern of world economic growth, which demands a scale of consumption that can be sustained only at the expense of continued deprivation and marginalization of the majority of the world, including women, from the processes and products of technology;

- A critical assessment of past limitations on technology, in both the developed and the developing countries, and its negative side effects on peace, self-reliance, social welfare objectives, and the equal integration of the rural and the urban poor and women into the national productive framework.

The United Nations Conference on Science and Technology for Development was a small but significant step in this direction. The Conference recognized that international market forces result in the transference of a form of technology that causes distortion and underutilization of resources (especially labour) and is incapable of creating an endogenous dynamic in developing economies. Different Conference documents examined the legacy of colonialism in the present pattern of technology transfer, the deleterious effects of colonialism on the status of women, and the technological choice appropriate to closing the gap between women's and men's labour productivity.[44] Thus the subject of the Conference afforded many opportunities to raise varied aspects of these issues relating to women.

Research, discussion, and action oriented towards women must not be seen as isolated tasks, but must be directly linked to larger research, evaluation, issues, and actions if the relevance of the results is ever to be communicated to policymakers and programme analysts, and if effective programmes facilitating equal female participation in development are ever to be developed and implemented. The role of women should be assessed in every agenda item of every global conference that focuses on development questions. Moreover, mechanisms for attaining an equitable and active female part in national and international development should be delineated within each agenda item of every global conference.

In the current process of restructuring the international economic, social, and information orders, and the institutional arrangements thereof, the opportunity should be taken to end the narrow confinement of issues affecting women to social and humanitarian affairs. Full recognition should be given to the demonstrable fact that there is no sector, department, policy question, or operational problem in which women worldwide are not intrinsically involved. Limiting the discussion of women's roles and inputs to certain issues directly related to women's biological and social reproductive responsibilities contributes to the oppressive glorification of these responsibilities and institutionalizes the systematic exclusion of women from equal integration in the development process.

Notes

1. Women's work, without having a contractual organization to regulate the distribution of domestic duties, the working time devoted to each of them, or the amount of payment for these services, is not paid in the form of wages based on the time that is effectively spent on this production. See Heleieth I. B. Saffioti, "Women, Mode of Production and Social Formations," *Latin American Perspectives*, Issues 12 and 13 (Winter and Spring 1977), Vol. IV, Nos. 1 and 2, pp. 27–37.

2. For a discussion of the relationship between sexually determined social categories and social classes, see ibid.

3. Lourdes Arizpe, "Women in the Informal Labor Sector: The Case of Mexico City," *Signs*, Vol. 3, No. 1 (Autumn 1977), pp. 25–27; Norma Chinchilla, "Industrialism, Monopoly Capitalism and Women's Work in Guatemala," *Signs*, Vol. 3, No. 1 (Autumn 1977), pp. 38–56; Deniz Kandiyoti, "Sex Roles and Social Change: A Comparative Appraisal of Turkey's Women," *Signs*, Vol. 3, No. 1 (Autumn 1977), pp.57–73; Ann Stoler, "Class

Structure and Female Autonomy in Rural Java," *Signs*, Vol. 3, No. 1 (Autumn 1977), pp. 74–89.

4. Celia T. Castillo, *The Changing Role of Women in Rural Societies: A Summary of Trends and Issues*, Seminar Report No. 12 (New York: Agricultural Development Council, 1977), p. 2.

5. For discussion of the relationship between reproduction and production, see Saffioti, "Women, Mode of Production and Social Formation," pp. 27–37.

6. M. J. Herskovits, *Economic Anthropology* (New York: Alfred A. Knopf, 1952); Claudine and Claude Tardits, "Traditional Market Economy in South Dahomey," in Paul Bohannan and George Dalton (eds.), *Markets in Africa* (Evanston: Northwestern University Press, 1962), pp. 89–102; B. W. Hodder, "The Yoruba Rural Market," in Bohannan and Dalton, *Markets in Africa*, pp. 103–17; Gloria Marshall, *Women, Trade and the Yoruba Family*, Ph.D. dissertation, Columbia University, 1964; Astrid Nypan, *Market Trade: A Samply Survey of Market Traders in Accra*, University College of Ghana Business Series 2, 1960; Margaret Katzin, "The Jamaican Country Higgler," *Social and Economic Studies*, Vol. 8 (1959), pp. 421–440; Alfred Metraux, *Making a Living in the Marbial Valley*, Occasional Papers 10 (Haiti: UNESCO, 1957); Sol Tax, "Economy and Technology," in Sol Tax (ed.), *Heritage of Conquest* (Glencoe, Ill.: Free Press, 1952), pp. 43–75; *Penny Capitalism*, Smithsonian Distribution Institute of Social Anthropology, Publication No. 16, 1953.

7. Sydney W. Mintz, "The Origins of the Jamaican Internal Marketing System – Internal Market Systems as Mechanism of Social Articulation," in American Ethnological Society, *Proceedings of the Annual Spring Meetings* (Seattle, Washington: American Ethnological Society, 1959), pp. 20–30.

8. Eleanor Leacock, "Women, Development and Anthropological Facts and Fictions," *Latin American Perspectives*, Issues 12 and 13 (Winter and Spring 1977), Vol. IV, Nos. 1 and 2, pp. 8–17: p. 11. For anthropological studies that discuss attitudes and practices that indicate women's former status and persisting importance in societies around the world see Patricia Draper, "Kung Women: Contrasts in Sexual Egalitarianism in Foraging and Sedentary Contexts," in Rayna R. Reiter (ed.), *Towards an Anthropology of Women* (New York: Monthly Review Press, 1975), p. 78; Eleanor Leacock, "Introduction," in Frederick Engels, *Origin of the Family, Private Property, and the State* (New York: International Publishers, 1975), p. 78, "The Structure of Band Society," *Reviews in Anthropology*, 1 May 1974, pp. 212–221: p. 221, and "Class, Commodity and the Status of Women," in Ruby Rohrlick-Leavitt (ed.), *Women Cross-Culturally: Change and Challenge* (The Hague: Mouton, 1975), pp. 608–609; Annie M. D. Lebeuf, "The Role of Women in the Political Organization of African Societies," in Denise Paulme (ed.), *Women of Tropical Africa* (Berkeley: University of California Press, 1971); W. J. McGee, "The Seri Indians," *Annual Report of the Bureau of American Ethnology*, XVII (Washington, D.C.: Government Printing Of-

fice,. 1898), pp. 269, 274; Marvin K. Opler, "The Ute and Paiute Indians of the Great Basin Southern Rim," in Eleanor Leacock and Nancy Lurie (eds.), *North American Indians in Historical Perspective* (New York: Random House, 1971), p. 269; A. R. Radcliffe-Brown, *The Andaman Islanders* (New York: Free Press of Glencoe, 1964), pp. 47–48; D. P. Sinha, "The Birhors," in M. G. Bicchieri (ed.), *Hunters and Gatherers Today* (New York: Holt, Rinehart and Winston, 1972), p. 317; William J. Smole, *The Yanama Indians: A Cultural Geography* (Austin: University of Texas Press, 1976), pp. 14, 31–32, 70, 75, 220, 221, 226; G. T. Basden, *Among the Ibos of Nigeria* (New York: Barnes and Noble, 1966).

9. Ida Rousseau-Mukenge, "Conceptualization of African Women's Role in Development: A Search for New Directions," *Journal of International Affairs*, Vol. 30, No. 2 (Fall-Winter 1976-1977), pp. 266–288; p. 266.

10. See Ester Boserup, *Women's Role in Economic Development* (London: Allen and Unwin, 1970); Laurel Bossen, "Women in Modernizing Societies," *American Ethnologist* II (November 1975), pp. 587–601; June Nash, "Certain Aspects of Integration of Women in the Development Process: A Point of View," paper prepared for the United Nations World Conference on the International Women's Year, Mexico City, 19 June–2 July 1975; Dorothy Remy, "Underdevelopment and the Experience of Women: A Nigerian Case Study," in Reiter (ed.), *Towards an Anthropology of Women*; Anna Rubbo, "The Spread of Capitalism in Rural Colombia: Effects on Poor Women," in Reiter (ed.), *Towards an Anthropology of Women*.

11. I. Muller and K. Fleischer, "The Status of Women in GDR Agriculture," unpublished paper prepared for the Fourth World Congress for Rural Sociology, Torun, Poland, August 9–13, 1976.

12. Elisabeth J. Croll, "Female Solidarity Groups as a Power Base in Rural China," unpublished paper prepared for the Fourth World Congress for Rural Sociology, Torun, Poland, August 9–13, 1976. However, despite deliberate policies to upgrade the socioeconomic status of women in China, some studies observe that the actual status of rural women has witnessed little positive change and that there is a division of labour by sex, which results in women standing in a peripheral relation to the dominant economic mode of organization. Weinbaum observed that women constitute 85 percent of the labour force of "neighborhood factories," a sphere characterized by a low level of capital, high labour intensity, the absence of state funds, and a wage rate 30 percent below that provided by the state sector. She stated that in China, the collective sphere is still where urban women primarily find employment and where 80 to 90 percent of the workers in collectively owned factories are women. In addition, she found that women are engaged in the service sector of socialized production three times as much as in material goods production. Women's involvement in this sector allows the payment of low wages in a low-priority and sparsely funded sector. Women also dominate in self-financed collectives, which are characterized by the absence of state investment. For further discussion, see Batya Weinbaum, "Women in Transition to Socialism: Perspectives on the Chinese Case,"

Review of Radical Political Economics, Vol. 8 (1976), pp. 34–58. See also the paper by Marina Thorborg of the University of Uppsala, prepared for the Lund Research Policy Institute Symposium on Technology in Development: India and China, May 22–24, 1978.

13. Ingrid Palmer, "Rural Women and Basic Needs Approach to Development," *International Labour Review*, Vol. 115, No. 1 (January-February 1977), pp. 102–103.

14. Ibid., p. 103.

15. Ibid.

16. Ibid.

17. Saffioti, "Women, Mode of Production and Social Formations," p. 29.

18. Carmen D. Deere, "Rural Women's Subsistence Production in the Capitalist Periphery," *Review of Radical Political Economics*, Vol. 8 (1976), pp. 9–17.

19. June Nash, "Women in Development: Dependency and Exploitation," *Development and Change*, Vol. 8 (1977), pp. 161–182.

20. Deere, op. cit.

21. Nash, op. cit., p. 166.

22. Judith Blake, "The Changing Status of Women in Developed Countries," *Scientific American* 231 (September 1974), pp. 136–147: 140.

23. Ibid., p. 142.

24. Nash, op. cit., p. 166.

25. Evidence concerning Brazil supports the contention that there is a relationship between Brazil's capitalist-oriented development strategy, the utilization of capital intensive forms of technology, and the worsening status of women. On the contrary, the statistics on women in Cuba's labour force demonstrate expanded access to technical training and female integration into productive activity. For statistics on Cuba, see Margaret Randall, *Cuban Women Now: After 1974* (Toronto: Women's Press and Dumont Press Graphix, 1974), pp. 6, 10. See also Ester Boserup, *Women's Role in Economic Development* (New York: St. Martin's Press, 1971), p. 151. According to Marjorie King, tipping and polarization, two phenomena that exacerbate the distinction between women's work and men's work, reserve for men the social status and power that comes through possession of technological skill and specialized knowledge. The incorporation of women at all levels of the labour force in Cuba offers women new skills, new knowledge, and the concomitant status and power. For further discussion, see Marjorie King, "Cuba's Attack on Women's Second Shift 1974–1976," *Latin American Perspectives*, Issues 12 and 13 (Winter and Spring 1977), Vol. IV, Nos. 1 and 2, pp. 106–120: p. 113.

26. Vinod Vyasulu, "Science and Technology for Underdevelopment," *New Scientist*, 18 January 1979, pp. 183–185. These points were also discussed in a national seminar on "Technology Choice in the Indian Environment," held in October 1977 in Bangalore.

27. Castillo, *The Changing Role of Women in Rural Societies*, p. 2.

28. See Elsa M. Chaney and Marianne Shmink, "Las mujeres y la moder-

nizacion a accesso a la tecnologia," in Maria del Carmen Elu de Lenero (ed.), *La Mujer en America Latino* (Mexico: Sepsetentas Dalla Costa, Mariarosa, 1975), pp. 25-26.

29. See Palmer, "Rural Women and Basic Needs Approach," pp. 101-102. See also Jane Hanger and Jon Morris, "Women and the Household Economy," in Robert Chambers and Jon Morris (eds.), *Mwea: An Irrigated Rice Settlement in Kenya* (Munich: Welt Forum Verlag, 1973).

30. M. P. de Thé, *Participation féminine au développement rural dans la region de Bouaké* (Abidjan: Ministère du Plan, 1968), p. 83. See also Palmer, "Rural Women and Basic Needs Approach," p. 101.

31. United Nations Document E/1978/106, *A Study of the Interagency Programme for the United Nations Decade for Women, Equality, Development and Peace.*

32. International Labour Organisation, *Employment, Growth and Basic Needs: A One-World Problem. Report of the Director-General of the International Labour Office and Declaration of Principles and Programme of Action Adopted by the Conference.* Tripartite World Conference on Employment, Income Distribution, and Social Progress and the International Division of Labour, Geneva, 1976. 2d ed., 1978, p. 61.

33. Mrs. Janet Mondlane, Representative of Mozambique to the Conference on Technical Cooperation Among Developing Countries, Buenos Aires, August 1978.

34. United Nations Document A/CONF.81/16, *Report of the United Nations Conference on Science and Technology for Development*, Vienna, August 1979, pp. 116-117.

35. Ibid., p. 42.

36. Ibid., p. 48.

37. Ibid., p. 99.

38. United Nations Document A/CONF.81/PC.15/Add. 1 (28 Nov. 1978), *Regional Report of the ESCAP submitted to the 3rd Preparatory Committee Meeting of the 1979 United Nations Conference on Science and Technology for Development;* see also United Nations Document A/CONF.81/RP.2, *The Regional Paper submitted by ESCAP to the United Nations Conference on Science and Technology for Development.*

39. United Nations Document A/CONF.81/NP.58, *The National Paper Submitted by the Democratic Republic of the Sudan to the 1979 United Nations Conference on Science and Technology for Development,* p. 361.

40. United Nations Document A/CONF.81/PC.17/Add.1, *Regional Paper for Africa*, Third Session of the Preparatory Committee, January 1979; United Nations Document A/CONF.81/RP.4, *The Regional Paper of Africa Submitted to the 1979 United Nations Conference on Science and Technology for Development,* p. 62.

41. United Nations Document A/CONF.81/NP.62 and Add.1, *The National Paper of Norway Submitted to the 1979 United Nations Conference on Science and Technology for Development,* p. 6.

42. Ibid., p. 12.

43. United Nations Document A/CONF.81/NP.16, *The National Paper Submitted by Sweden to the 1979 United Nations Conference on Science and Technology for Development*, p. 65.

44. World Conference of the United Nations Decade for Women: Equality, Development and Peace, Document A/CONF.94/19, *Recommendations Relating to Women and Development Emerging From Conferences Held Under the Auspices of the United Nations or the Specialized Agencies, Item 9(b) of the Provisional Agenda*, p. 30, paragraph 87–89 (Report prepared for the Department of International Economic and Social Affairs by Ingrid Palmer, Consultant).

6
International Policymaking
for Development Through
the United Nations:
Science and Technology
Before and After UNCSTD

Volker Rittberger

1. UNCSTD: A Post-Mortem

The United Nations Conference on Science and Technology for Development (UNCSTD), which took place from 20 August till the early hours of 1 September 1979 in Vienna, focused international attention on a key area of policymaking for development. Widely differing as the views of experts and practitioners about development and the role of science and technology in shaping its course may be, few would dispute the salience of this policy field and the need for international action on it. The expectations that this Conference aroused are summed up in the opening statement by the Secretary-General of the United Nations, Kurt Waldheim, who stated

> that the Conference was the latest in a series sponsored by the United Nations to help find answers to the growing problems of the modern era. In one way or another, the entire spectrum of global ills related to the creative uses of science and technology in development. Much of human ingenuity and innovative ability had been misdirected, for example, into the refinement of military technologies and into the support of wasteful consumerism in a world where famine and malnutrition were tragically present. It was a major task of the Conference to help ensure that scientific and technological potential should be directed to constructive ends. There was an enormous imbalance in the research and development activities being undertaken throughout the world, in that 97 percent of such activities took place in the industrialized countries. The developing countries, as a whole, were excessively dependent on imported technologies, which tended to hamper the growth of indigenous skills. The developing countries needed help in order to gather and share scientific knowledge, so as to enhance their technological capabilities and accelerate their development. This required a global programme of action which the Conference was designed to evolve.[1]

In formulating these expectations toward UNCSTD the Secretary-General of the United Nations was seconded by the Secretary-General of the Conference, Joao Frank da Costa, who pointed out that

> the Conference had already in the preparatory phase achieved one of its objectives, since the importance and specificity of science and technology, as instruments of development, had been broadly recognized at the national level. Such a recognition should now be manifested at the international level. The Conference should be re-

garded not merely as a means of arousing public awareness, but as a global gathering which should culminate in positive recommendations for action at national and international levels.[2]

He added that

the three critical issues before the Conference were the transfer of technology, the institutional mechanisms, and the financial problems.[3]

After elaborating on these issues he concluded with the observation that

the Conference was taking place at a turning point in history and in a sense marked the end of one era and the beginning of another. The international community should try to identify in the course of the Conference the interests common to developed and developing countries, to Governments and private industry and to producers and users of technology. The results of the Conference would show that the real division of the world was not between developed and developing countries but between those who wanted to preserve the *status quo* and those in favour of the individual and collective development of all countries in an atmosphere of innovation and change.[4]

As against these publicly professed, far-reaching expectations of the United Nations executive leadership, post-Conference assessments and evaluations by participants and observers cover a wide range, from decrying the Conference as a failure to acknowledging its achievements, limited as they may be. Except for the Secretary-General of the Conference, who declared that "practically all sectors agree that the Vienna Conference was a success — except in a few isolated instances due to lack of information or lack of understanding of the problems,"[5] there seems to be general agreement that UNCSTD represented neither a landmark event in the North-South dialogue nor a breakthrough in international policymaking for development in the field of science and technology. In a survey of how a variety of actors involved in the Conference, or concerned about its results, viewed the outcome of UNCSTD, Klaus-Heinrich Standke, former Director of the United Nations Office for Science and Technology, reached the following conclusion:

The visible results of UNCSTD are relatively modest: a new committee machinery, a new secretariat structure and a voluntary fund. . . .

Nevertheless, readiness to reorient the role of science and technology in the complex process of development was achieved through the Vienna Conference. Even if the desired interaction between politicians on one side and scientists and development planners on the other side did not take place during the Conference itself, the new organizational mechanism within the United Nations system, coupled with the anticipated new financial means, could allow for new worldwide initiatives.

In any event, UNCSTD was a qualified success; having—through the mechanism of a World Conference—reached agreement on the Vienna Programme of Action to which all participating government delegations were able to subscribe, is an encouraging achievement. This political victory which, in a world full of controversies, antagonisms and confrontations, should by no means be played down, was found to have its price on the substance of the Conference outcome.[6]

However, Standke failed to be explicit about the nature and the amount of the alleged "price on the substance of the Conference outcome" that the compromise solutions reached at the Vienna Conference exacted. Moreover, although he did not take refuge in the simplistic formula of "lack of political will" to explain the Conference outcome, he had little to say about the interest configuration prevailing among and within the Member States as well as the member organizations of the United Nations system that were involved in the Conference process. There can be no doubt that the analysis of this interest configuration is a key to placing this and other global ad hoc conferences of the United Nations system in proper perspective.

2. United Nations Global Conference Diplomacy and Policymaking in the International System

A. *Global Conference Diplomacy and Policy Coordination*

The type of international deliberation and decision-making procedure of the United Nations system that is represented by UNCSTD, i.e., problem-oriented global conference diplomacy, may be viewed as a distinctly novel element in the organized political relations among nations, and especially between developing and developed countries. Norman Graham and Stephan Haggard have noted that conference diplomacy has, of course, been an integral part of international organization since the Congress of Vienna, but the emergence of international bureaucracies in the twentieth century has changed its character significantly.

The current proliferation of "special sessions" and *ad hoc* conferences on specific topics—from disarmament to primary health care—are indicative of a crisis in international organizations. . . . Among other things, they [the "special sessions" and *ad hoc* conferences] underline gaps in the programming of organizations, areas where national leaders feel constrained, for whatever reasons, to seek international solutions. They also provide a spotlight for populist reform movements and debates over the policy and direction of the organizations involved.[7]

Since the beginning of the 1970s in particular, the United Nations system has resorted with increased frequency to this procedure. During this decade, the United Nations system organized or sponsored about two dozen special sessions, ad hoc world conferences and special negotiating conferences dealing with almost every conceivable issue of international development and the New International Economic Order. The frequency with which global conference diplomacy has been used can be assumed to be indicative of a widespread interest in international policy-coordination both at the interstate level and at the interorganization level. At the same time, however, it can also be assumed that this interest in international policy-coordination varies from one issue area to another. Nevertheless, as there are both objective, functionalistic and subjective, strategic or tactical explanations for this interest, the process and structure of international policy-coordination deserves closer scrutiny.

First of all, the concept of international policy-coordination refers to a multilevel decision-making system that is expected to cope with the pressure of problems resulting from structural deficiencies in the present international system and from performance deficiencies caused by the discrepancy between problem space and action space at the level of an individual decision-making unit, e.g., the central government of a nation-state. Although the multilevel decision-making system of international policy-coordination transcends the fragmented decentralization of international policymaking by nation-states, it differs from a centralistic, supranational decision-making system in that national decision-making units remain necessary participants in the overall system.[8]

Second, the objective, functionalistic interpretation of an interest

in international policy-coordination, especially as related to issues
of international development and the New International Economic
Order, rests on the almost universal recognition that the
underdevelopment of the Third World and the asymmetric
dependence of most developing countries on the industrially
developed countries has created a host of pressing problems for
either group of countries as well as for both groups together. There
is unanimity at least on the minimalist demand that the widening
of the gap between industrially developed and developing countries
(and between the rich and the poor) must be halted and preferably
reversed. Disagreements arise, however, as soon as the question of
devising an operational strategy of change is raised, including the
issue of whether and how a systematic interlinking of international
and intrasocietal structural change should be brought about.

Faced with this global development problematique, can we rely
on a decentralized, nationally fragmented international policymaking system to deal with the many issues deriving from it? The experiences of the recent past as well as current trends seem to indicate that the answer is likely to be negative. For, indeed, the costs
imposed on world society as a whole, but also on most developing
and industrially developed countries individually, by the pursuance
of autarchic or semi-autarchic development strategies would probably be heavy enough to demonstrate clearly the suboptimal nature
of a purely decentralized international policymaking system. The
disadvantages of purely bilateral economic and development
cooperation between developing countries and industrial states
would seem to point in the same direction.[9]

Third, a noncentralistic alternative to the system of nationally
fragmented international policymaking would seem to be compatible with, if not preferable in view of, the subjective, strategic interests of a majority of the political and administrative actors in the
international system. They can be assumed to reckon that it will
bring them relief from the pressure of problems, both those emanat-
ing from their own social environment and those attributable to
other political and administrative actors. We can assume further
that a majority of political and administrative actors perceive the
multilevel decision-making system of international policy-coordination as an opportunity to achieve easier external acceptance
of programmes and internal control of demands, but also as a kind of
scapegoat in the event that self-set objectives are not attained. The
vested interests of international (development) bureaucracies in the
further evolution of this system derive from the fact that their existence and competence to take action both derive from and are cir-

cumscribed by international policy-coordination: Clearly, they do not constitute a self-propelling force toward a centralistic, supranational system of international policymaking.

A centralistic response to the challenge of international policymaking is out of the question, at least for the foreseeable future,[10] so international policy-coordination in a multilevel decision-making system suggests itself as the only viable alternative to nationally fragmented policymaking in the international system. This multilevel decision-making system encompasses both governments of nation-states and nonstate multinational policymaking institutions. The latter either facilitate horizontal coordination or are directly involved in coordination through joint decision making. Put differently, horizontal coordination takes place when the decentralized decision-making units that are in need of coordination participate in the interaction themselves, assisted by nonstate multinational agencies. On the other hand, we can speak of coordination through joint decision making when nonstate multinational units are necessary participants in policymaking processes in which states, however, are not subordinated to the former but keep control over them in all of their major phases.[11] These nonstate multinational policymaking institutions are what are commonly but mistakenly referred to as "intergovernmental organizations" (IGOs). The United Nations system is composed of institutions that may be said to be involved primarily in horizontal coordination, whereas the European Community is representative of institutions that are part of a system of international policy-coordination through joint decision making.

The viability of international policy-coordination as a noncentralistic alternative to nationally fragmented international policymaking rests on its capacity to evoke positive answers to the following questions:

1. Does it attain a level of information gathering and processing that is superior to the decentralized system of decision making through nation-states?

2. Does it have the capacity of reconciling diverging, if not conflicting, interests so that consensual and practicable solutions with high problem-solving effectiveness can be found?

3. Does it muster the collective resources and maintain the collective commitment to ensure adequate implementation of the decisions taken?

B. A Theoretical Digression

Multilevel policymaking systems are faced with the task of giving due consideration to two different sets of constituencies. On the one hand, there are the needs and demands, but also the resources and action potential, of people, groups, and organizations that affect, and are affected by, public policymaking. On the other hand, the participating politico-administrative units partly act as brokers and filters of the above inputs and partly inject their own institutional interests into the policymaking process. The analysis of public policymaking in pluralistic and federal states has given rise to two different paradigms that have considerable heuristic value for the study of international policy-coordination. The "neocorporatism" paradigm[12] focuses on policy coordination between governments and key interest groups, whereas the "policy coordination" paradigm in its narrower sense[13] highlights interorganizational decision making involving more than two politico-administrative units at the same level of government and/or at different levels of government.

If we turn to the analysis of public policymaking in the international system involving multinational politico-administrative units, such as institutions of the United Nations system or of the European Community, it may be said that the neocorporatism paradigm tends to converge with the policy coordination paradigm. The main reason is that the representation of organized social interests in the international system independent of state institutions has not evolved very far; rather, it is the national states and their agents through which social interests, more often than not, gain access to, or are shut out from, the arenas of international public policymaking. Thus, the neocorporatism paradigm could provide only an analogy for studying processes of international policy-coordination, whereas the policy coordination paradigm in its narrower sense is, with appropriate adaptation, directly applicable and, thus, preferable for the purposes of the present analysis.[14]

However, the policy coordination paradigm, at least at first sight, imposes certain restrictions on the scope of international public policy analysis. It tends to filter out social interests and problem perceptions other than those deemed acceptable and articulated by the agents and institutions of national states. Thus, this paradigm may be courting the risk of structurally ignoring policy alternatives that may, or may not, be superior to those actually falling within its analytic purview. Yet, one might argue that the large number of national states itself creates a high probability that a wide range of social interests and problem perceptions will be represented in international arenas of public policymaking. Therefore, one might expect to find the major policy alternatives in a given issue area being articulated and legitimized despite the particularism and selectivity of interest representation demonstrated by individual nation-states in international settings.

3. International Policy-Coordination in the Field of Science and Technology Before UNCSTD

Science and technology number among those fields in which the necessity of organizing the satisfaction of certain social needs as a joint international task[15] was recognized and accepted at a very early stage. Referring to the nineteenth century and its second half in particular, Brigitte Schroeder-Gudehus noted that

> governments became increasingly concerned with scientific and technical matters specifically linked to public services (i.e., meteorology, epidemiology, seismology, etc.). Indeed, a considerable amount of intergovernmental cooperation developed in functional areas where the availability of new knowledge, techniques and instruments required collaboration: coordination of activities, exchange of information, standardization of measurements, etc. International committees or central bureaus were set up in many instances. These bodies were usually linked to the corresponding technical departments or agencies of the member states, departments which also appointed and instructed the delegates. With a few exceptions, such as the Meter Conference of 1875, this cooperation scarcely caught the attention of high level diplomats. The ministries of foreign affairs were involved generally in only formal aspects: drafting of agreements, signatures, annual payment of membership fees, etc.[16]

Even though joint international policymaking on the matters referred to above still occupies a prominent place in international scientific and technological affairs, with the inclusion of the developing countries partly achieved through the mechanisms of technical assistance and advisory services, the concerns of international policymaking in the field of science and technology, and the problem perceptions that have prompted them, have become much broader and certainly more politicized. There exists now a wide range, both substantively and spatially, of cooperative endeavours to promote scientific and technological change, to make use of its results, and (a still underdeveloped area) to assess and control its consequences.[17] Yet, the vast numbers of cooperative endeavours in science and technology and their distribution across nations have

actually sharpened the perception of one of the most crucial
political problems in the field of science and technology: the
dependence of the Third World on the industrially developed countries. In the following, therefore, the genesis of joint international
tasks of global scope stimulated by an evolving problem perception
of scientific and technological dependence will be singled out to examine the motive forces, functions, and limitations of international ·
policy-coordination in a given issue area and to highlight the potentially catalytic role of ad hoc global conference diplomacy.

Ever since the formal proclamation of the First Development
Decade of the United Nations[18] the narrowing of the socioeconomic
gaps between developing and industrially developed countries has
been regarded as a joint international task of topmost rank. The
salience of this task has been reinforced with the consensus reached
on the International Development Strategy for the Second Development Decade of the United Nations.[19] The resolutions and programmes of action adopted at the sixth and seventh Special Sessions
of the United Nations General Assembly[20] underlined the urgency
of moving ahead with this joint international task. The International Development Strategy agreed upon at the Eleventh Special
Session of the United Nations General Assembly[21] amidst the
unresolved conflict over the proposed round of global negotiations
on key development issues has again confirmed the status and the
legitimacy of the task, but may have been inspired by a too optimistic outlook on overcoming the present world economic
crisis.

Within the framework of the broadly defined joint international
task of overcoming the underdevelopment and dependence of the
developing countries, science and technology have been assigned
the role of key variables in the process of social, economic, and
cultural development. However, in the United Nations General
Assembly's resolution on the First Development Decade[22] science
and technology were given anything but an outstanding role. In
subsections 4(e) and 4(f) of this resolution, only a general reference
was made to scientific and technical training needs and to the
utilization of the potential inherent in science and technology for
the promotion of development. Subsequently, the then Secretary-
General of the United Nations, U Thant, convened a United Nations Conference on the Application of Science and Technology for
the Benefit of the Less Developed Areas (UNCSAT), which was held
in 1963 in Geneva, but the general assessment was that it had failed
to produce any politically significant results.[23] In particular, the Con-

ference added little, if anything, to the formulation, let alone to the implementation, of the joint international task of strengthening the science and technology sectors in Third World societies. Moreover, the institutions within the United Nations set up in the aftermath of UNCSAT—the Office for Science and Technology and the Advisory Committee on the Application of Science and Technology (ACAST)—needed time to become operational, but then contributed heavily to the formulation of the Strategy for the Second Development Decade.[24]

A marked change occurred with the adoption of the International Development Strategy for the Second Development Decade in 1970. In the United Nations General Assembly resolution, a special section was devoted to international cooperation for development in the field of science and technology. Clearly, the provisions of this section went far beyond the relevant sections of the resolution on the First Development Decade as regards the implied problem perception, the stated strategic objective, and the proposed measures.[25] Furthermore, a permanent Committee on Science and Technology for Development (CSTD) of the United Nations Economic and Social Council was established in 1972.[26] But, essentially, things were taken no further than the more precise and more differentiated conceptualization of the new joint international task within the framework of the United Nations system. Even the beginnings of an operational planning, regulatory, and financing system involving the Member States and organizations of the United Nations system remained far off. Nor did any basic new initiative emerge from the sixth and seventh Special Sessions of the United Nations General Assembly. However, the Seventh Special Session made another successful contribution to sharpening the conceptualization of the joint international task(s) in the field of "science and technology for development," as it had become known since the inception of the Second Development Decade. Over and above this, it lent its support to the recommendation made by an Intergovernmental Working Group of the CSTD to hold a special conference of the United Nations on science and technology for development before the end of the decade, i.e., before 1980.[27] The formal resolutions on the convening and the mandate of such an ad hoc world conference were adopted during the sixty-first session of the United Nations Economic and Social Council[28] and the thirty-first session of the United Nations General Assembly.[29]

The motive forces behind these resolutions were first and foremost the demands put forward by the developing countries organized in the Group of 77 within the framework of the United Nations. These demands aimed at achieving at long last measurable progress in closing the scientific and technological gap between the North and the South and in reducing the resulting dependence of the Third World. In addition, the conference idea has found support in most industrialized countries. There, the perceptions of the Conference purpose were sometimes diffuse, sometimes even contradictory. Basically, decision makers approved of, and wanted to support, the strengthening of the scientific and technological potential of the developing countries; yet, they felt no inclination to forego the safeguarding of long-standing competitive advantages in the very field of science and technology. Thus, their apparent commitment to change was more abstract than real. Lastly, there were also the partly converging, partly diverging vested interests of various member organizations of the United Nations system and their subunits specializing in science and technology affairs. Although they generally favoured the proposal of holding a special conference, expecting it to strengthen their competence in this field, some seemed to fear at the same time the possibility of new or increased interorganizational competition for scarce resources in the face of selectively broadened mandates.[30]

After this brief review of trends in international policy-coordination involving the United Nations system before UNCSTD we shall now proceed to examine whether, to what extent, and how UNCSTD has contributed to the emergence of an operational planning, regulatory, and financing system at the global level. The first step consists of determining and assessing what UNCSTD (and its preparatory process) has accomplished with respect to information gathering and processing. Put differently, we are interested here in the quantity and, above all, quality of its performance in providing action-oriented analyses of the crucial problems in the field of science and technology for development.

4. UNCSTD as a Catalyst of Information Processing and Problem Identification

A. Selection of Information Sources

If there was anything special about the preparation for UNCSTD, it was the deliberate attempt to organize UNCSTD policymaking as an "ascending process": The preparatory period, conceived as "an integrated and fundamental component of the Conference itself,"[31] was to generate a forward-looking, action-oriented programme of action for the development of science and technology, particularly in developing countries, which would be based on information and analyses submitted by political and administrative units at the national, subregional, and regional levels. Both before the final resolution was adopted on the holding of UNCSTD[32] and again very intensively during the first session of the Preparatory Committee for the Conference (31 January–2 February 1977, New York),[33] the most comprehensive and detailed stocktaking possible of national experiences with the role of science and technology in the process of national development was advocated, especially by the representatives of some Latin American countries and the newly appointed Secretary-General of the Conference, the Brazilian Joao Frank da Costa.[34] They also gained acceptance of the view that such national experience should be surveyed by studies undertaken by, or under the authority of, the competent political and administrative organs and agencies of the member countries. Conversely, this meant that, in contrast to a number of earlier ad hoc world conferences of the United Nations system, the Conference documentation should be based to as small a degree as possible on the contribution of nongovernmental experts ("wise men").[35] Accordingly, the Member States were requested to submit national papers within a certain period of time, which, as a consequence of both the novelty of the task and the lack of momentum in the beginning of the preparatory period, had to be extended considerably. Thus, the time available for a thorough analysis of their contents, on which the Conference documentation was to be based, was drastically reduced.

The adoption of guidelines for drafting the national papers[36] was preceded by lengthy discussions and required a considerable amount of bargaining.[37] On the one hand, the desirability of issuing very detailed guidelines was questioned. On the other, and more importantly, representatives of many industrial countries had misgivings about too broad a definition of the mandate for the Conference. Therefore, they proposed the selection of a limited number of problem areas of applying science and technology to development on which the work of the Conference should be concentrated. These proposals were motivated by the desire to "depoliticize" the Conference and, in particular, to delink it from the New International Economic Order. As against this approach, the developing countries, led by some of the more advanced ones, persisted in providing a definition of the mandate for the Conference that focused on the scientific-technological gap between North and South and on the dependence of the developing countries and, thus, put it squarely within the context of the New International Economic Order. In the end, the developing countries prevailed with their definition of the mandate for the Conference as well as with their notions about the guidelines for national papers, thus determining the direction that the process of information gathering was to take. As a concession to the industrially developed countries' concerns mentioned above, it was agreed that five "subject areas" would be selected, but for illustrative purposes only. After the first round of regional consultations during 1977, the Preparatory Committee, at its second session (23 January–3 February 1978, Geneva), established a list of five subject areas;[38] yet, during the remainder of UNCSTD they went by and large unnoticed. Instead, they reappeared in the agenda of the Scientific Colloquium that was organized by ACAST and that preceded UNCSTD in Vienna by one week.[39]

In addition to requesting the submission of national papers, the Preparatory Committee for UNCSTD also decided that a series of regional consultations should be held with the participation of the five regional commissions of the United Nations. These regional consultations were intended to result in regional papers also to be submitted to the Preparatory Committee.[40] Originally, the Secretary-General of the Conference and the majority of the Preparatory Committee envisaged the drafting of the national and regional papers as a two-step process, with the regional papers already achieving a kind of synthesis of the national papers from their respective regions. However, the substantial delays in the initial phase of Conference preparations made it necessary to revise this conception and to settle for a more or less parallel process of preparing both the national and the regional papers. In each region at least two rounds of formal consultations were held, first at the level of governmental experts (1977) and then at the ministerial level (1978). In some instances, regional consultations were preceded by subregional meetings, and in the case of the Latin American region it proved necessary to convene a third meeting to approve the regional paper.

In line with the stated objective of organizing information gathering and processing during the preparatory period of the Conference as an "ascending process," the formal submission of topical papers by the organizations of the United Nations system, other intergovernmental organizations, and nongovernmental organizations was not called for. These organizations were asked only to submit "background papers" for UNCSTD itself, i.e., at a time when the work on the analytical Conference documentation for the agenda items had been completed and when the Programme of Action would be in an advanced stage of drafting.[41] However, the organizations of the United Nations system were entrusted with preparing two collective papers to be submitted to the Preparatory Committee, which enabled them to bring to the attention of the Committee information that they deemed essential from their respective institutional points of view. One paper was to give an overview of the mandates and activities (past and present) of the organs, bodies, and organizations of the United Nations system in the field of science and technology for development; the other paper was to review the implementation of recommendations adopted by preceding conferences of the United Nations system that related to science and technology for development.[42] As the preparatory period drew to a close, the organizations of the United Nations system were asked to collaborate on two cross-organizational analyses pertaining to programme activities and to available resources in the field of science and technology for development; the short time available for their drafting was bound to limit their usefulness.[43] To round out the picture of United Nations system involvement in the preparatory process, it should be added that some organizations, including UNESCO, FAO, and UNCTAD, had seconded officers to the secretariat of UNCSTD through which they maintained a continuous two-way flow of information.

Nongovernmental organizations in general and the scientific and technological communities in particular were largely excluded from providing direct information inputs for the preparatory work of the Conference. Instead they were advised to get involved in the national preparations for the Conference to have their concerns taken into consideration.[44] Through the exclusion of the nongovernmental organizations and the scientific and technological communities one important function of the "ascending process" was highlighted, which was to shield the delegates of developing countries in particular from any transnational consensus that scientists, intellectuals, and other concerned nongovernmental groups might reach on questions relating to science and technology for development. Consistent with this approach the roles of ACAST and of the Office for Science and Technology within the United Nations secretariat were downgraded as, traditionally, they had close connections with these groups. The demise of these two institutions during the preparatory process was facilitated politically by two inherent defects: One was the disproportionately high representation of industrially developed

countries' nationals in these institutions; the other had to do with
the (varying degrees of) aloofness of developing countries' nationals
in these institutions from the power centers in the Third World.

B. Problems of Information Processing

After it had been established which sources were to supply the
relevant information for conference policymaking it remained to be
seen whether and how information gathering and processing by
those actually participating in conference policymaking would rely
on, and make use of, these sources. For all their quantitative and
qualitative differences, the national papers undoubtedly constitute
a formidable source of information on the substantive items of the
conference agenda. At the time of the opening of UNCSTD more
than 120 countries and recognized national liberation movements
had submitted their national papers.[45] At the originally set deadline
for submission, 1 May 1978, however, only two papers had been
received by the Conference Secretariat in their final form and twelve
papers in a provisional version. Even on 1 August 1978, the
extended deadline for submission, the Conference Secretariat had
not more than forty-seven papers in their final form and another
thirty-six papers in a provisional form.[46]

The Conference Secretariat, which was entrusted with analyzing
and digesting the information contained in the national papers for
the preparation of the conference documentation, found itself confronted with a task that was already difficult to perform under the
existing constraints of time and personnel and was exacerbated by
the irregular inflow of national papers. The outline of a programme
of action submitted by the Conference Secretariat to the General
Assembly in October 1978 comprised essentially a compilation of
more than 200 proposals for action broken down in accordance with
the substantive items of the Conference agenda.[47] The General
Assembly then called on the Conference Secretariat to prepare for the
third session of the Preparatory Committee (22 January–2 February
1979, New York) a more analytically arranged, but nevertheless
action-oriented, provisional draft programme of action.[48] Although
the Conference Secretariat succeeded in complying with this request
under heavy pressure of time and submitted such a draft programme
of action setting forth six target areas,[49] it drew mixed responses.
The Group of 77 demonstrated little enthusiasm, whereas the industrially developed countries from both East and West expressed
some reservations and indicated possible revisions, but made it
clear that, in general, they were prepared to accept the Secretariat's

document as a basis for further consultations. At the request of the Group of 77 the Preparatory Committee decided to return this provisional draft to the Conference Secretariat but failed to provide any precise directives for revising it.[50] For the fourth session of the Preparatory Committee (23 April–4 May 1979, New York) the Conference Secretariat submitted yet another draft in which it attempted to adhere to the wishes vaguely indicated by the spokesman of the Group of 77 (Tunisia) during the third session.[51] However, the Secretariat's draft was superseded by an initiative of the Group of 77 itself and played no part in the further consultations on the Programme of Action. So as early as spring 1979, the project of a thoroughly structured "ascending process" of information gathering and processing for Conference policymaking had reached a dead end. Moreover, the function of the Conference Secretariat as supplier of a raw synthesis of information furnished by national and regional papers and covering both the analytical and action-oriented parts of the Conference documentation had become obsolete.

The failure of the "ascending process" was further dramatized by the fact that, in effect, the preparation of the principal action-oriented Conference document, i.e., the Programme of Action, was separated from the elaboration of the analytical Conference documentation, i.e., the inventory and analysis of problems relating to the substantive items of the Conference agenda. It was left to the Conference Secretariat to continue with the work on the analytical Conference documentation, the volume of which was drastically reduced by the Preparatory Committee at its third session.[52] On the other hand, the further elaboration of the programme of action was taken over *de facto* by the Group of 77 on its own when, in early 1979 and after the third session of the Preparatory Committee, it set up a working group in New York. Making use to some extent of the resources of the Conference Secretariat, but being more effectively screened off than the Secretariat from attempts by industrially developed countries and United Nations organizations to influence its work, the working group of the Group of 77 produced a draft for a programme of action that gained the status of a basic document for consultation on the Programme of Action during the fourth and fifth sessions of the Preparatory Committee.[53]

Hence the national papers, and likewise the regional papers, were, by and large, dispensed with as a direct source of information for conference policymaking. An important reason for the reversal of the developing countries' position on the role of the national and regional papers may have been the fear among some of them that the complex nature of the information emerging from the papers as well as information overload might detract from pursuing the foremost objective of UNCSTD: the restructuring of North-South relations in the field of science and technology. Since the Conference Secretariat had to act as custodian of that fund of information represented by the national and regional papers, it was only logical for the developing countries to minimize the substantive role of the Conference Secretariat during the decisive phases of consultations on the Programme of Action. With a lightened burden of information, the Group of 77 concentrated on reaching a group consensus on a set of issues (policy objectives, legal framework, and institutional and financial arrangements) that reflected their chief political interest in the Conference. Only after having worked out compromises among themselves were negotiations on a mutually agreeable text opened with the industrially developed countries, which neither individually nor collectively had shown an inclination to present their own version of a draft programme of action.

C. Identification and Clarification of the Main Policy Issues

Information gathering and processing during the Conference process was influenced by political definitions of the central problems in the field of science and technology for development. These definitions were given different degrees of preference by various groups of countries. Following the categorization developed in Chapter 1, Section 5, we distinguish three approaches to defining the main policy issues that centre around the concepts of dependence (*dependencia*), global problems, and social control. To summarize, the first of these approaches focuses on the closing of the scientific and technological gap between North and South, which is considered as primarily due to external causes, and on overcoming the scientific and technological dependence of the developing countries on the industrialized countries. The second approach stresses chiefly the worldwide inadequacy of scientific and technological problem-solving capabilities and perceives their augmentation to be the key to cope better with the global development problematique. The third approach, finally, draws attention to the social functions of scientific and technological change and to the consequences of uncontrolled change for socioeconomic and political development both within and between nations.[54]

On reviewing the Conference process and the Conference documentation, it would appear admissible to conclude that the *dependencia* approach to defining the main policy issues in the field of science and technology for development clearly dominated information processing during the decisive stages of conference policymaking.[55] Most of the leadership groups in the Third World have espoused development nationalism in one form or another, and the *dependencia* analysis of underdevelopment has been one of its crucial theoretical underpinnings. Its ideological appeal derives from the emphasis it places on the exogenous causes and conditions of underdevelopment and from its ability to identify the actors in this historical process. However, as development nationalism in practice tends to blend conflictive and cooperative orientations toward the industrially developed countries,[56] the *dependencia* approach actually leads to a double-faceted problem definition: Science and technology for development constitute both a problem of increasing the scientifiic and technological capabilities in general and of developing countries in particular and a problem of distributing the scientific and technological potential equitably among nations. Irrespective of the sometimes strongly worded demands for change arising from the adoption of the *dependencia* approach, it is more adequately characterized by its conflict-minimizing function, which results from the expectation implicit in this approach that growth and redistribution are interacting and mutually supportive processes. Accordingly, the role of multinational political and administrative institutions is, above all, to help coordinate policies for growth and identify appropriate policy instruments for this purpose. Second, they are to foster agreement on allocating external resources to states that lack a modicum of appropriate policy instruments (including financial resources) for stimulating growth in the field of science and technology. Obviously, it is this task that often engenders, particularly among net donor states, strong resistance to what they perceive is international policy-coordination going too far.

The dominance of the *dependencia* approach, however, does not exclude the possibility that elite segments in the Third World may share the view, which has a very strong backing in all relevant strata of the industrially developed countries, that decisions relating to science and technology for development should be guided by the global problems approach. This approach has also found endorsement in the Conference documentation, albeit less prominently and certainly subordinated to the *dependencia* approach.[57] The global problems approach clearly lacked strong support among the diplomatic representatives of the developing countries. This must be attributed to the nature of the problem definition to which this approach lends itself: The collective management of science and technology for development, at both the national and the international level, is seen primarily as the setting of global standards and goals and the mobilization of the social efforts and material resources necessary to attain these goals or to comply with these standards. Thus, science and technology appear to be distribution-indifferent instruments for social problem solving. This approach also implies that it does not matter who performs the problem solving for whom. The emphasis on international cooperation inherent in this approach is very strong but carries an unmistakable bias for scientifically and technologically potent partners in cooperation.

Compared to the *dependencia* approach, and even the global problems approach, the social control approach plays at best a marginal role in the Conference documentation.[58] Had it not been for the representatives from the Nordic countries this approach would have gone largely unnoticed by UNCSTD; and even for them, the social control approach had but a very relative weight. It was given a generally poor reception, particularly by developing countries' representatives, because it presented a view of setting standards and of tackling patterns of maldistribution that implied a strong criticism of existing social and political practices in a very large number of countries. The possibilities, highlighted by this approach, of the abuse and injurious consequences of science and technology, and the recognition of disparities in the use of and access to science and technology within nations stand in sharp contrast with the basically optimistic perspective on the role of science and technology in development that is characteristic of both the *dependencia* and the global problems approaches. The obstacles to accepting this approach more readily are strengthened by the fact that its implications for collective international action would seem to represent a far-reaching encroachment on national sovereignty.

Information processing as well as problem definitions and analyses undertaken in the course of the conference process pointed out — though for varying reasons — the necessity of enlarging and intensifying international policy-coordination in the policy field covered by UNCSTD.[59] However, a decision on setting up an operational planning, regulatory, and financing system for science and technology within the framework of the United Nations was not a foregone conclusion. Thus, in a further step, we will examine the Conference results in greater detail and take account of how the various interests represented in the Conference process succeeded in being incorporated into, or in blocking, consensus formation.

5. UNCSTD's Contribution to Goal Setting and Programme Formulation

A. *The Vienna Programme of Action*

The Vienna Programme of Action for Science and Technology for Development is certainly the most conspicuous result of the Conference policymaking process. Thus, it seems appropriate to begin this section with a summary of the main propositions of the Vienna Programme of Action.

The Programme of Action is divided into three parts, each corresponding to a set of action proposals oriented toward a paramount programme objective ("target area"):

I.   Strengthening the Science and Technology Capacities of the Developing Countries.

II. Restructuring the Existing Pattern of International Scientific and Technological Relations.

III. Strengthening the Role of the United Nations System in the Field of Science and Technology and the Provision of Increased Financial Resources.[60]

In the first target area, the Vienna Programme of Action calls on the developing countries to formulate a national science and technology policy that aims at the development of both suitable policy instruments and the broadest possible range of scientific and technological capacities "in a country-specific, resource-specific and product-specific pattern" (paragraph 19). In all developing countries, specific rules should be laid down and effective procedures established for managing and appraising the transfer and acquisition of technologies. The developing countries are further urged to give priority consideration to the basic and advanced training of specialists and management personnel as well as to the problem of the so-called brain drain, to promote the close intermeshing of research and development on the one hand and of the production and distribution systems on the other, and to support the expansion of the scientific and technological information and extension services.

The Programme of Action also enjoins the industrially developed countries to provide generous assistance to the developing countries for the purpose of strengthening their science and technology capacities. The Vienna Programme enumerates a number of suitable supportive measures to be adopted and carried out by the industrially developed countries: increasing the support for research cooperation with developing countries involving a rapidly growing number of scientists and technologists from these countries; providing systematically for the transfer of results from R&D work and innovation; raising the level of R&D expenditures and activities for solving crucial problems of developing countries representing obstacles to rapid development ("bottlenecks"). In order to strengthen the scientific and technological manpower potential of the developing countries, the Programme of Action strongly recommends that the industrially developed countries should provide more scholarships both for studies in industrially developed countries and for basic and advanced training programmes in the developing countries themselves.

However, even in this first target area, the Vienna Programme leaves no doubt that, irrespective of the assistance expected to come from the industrially developed countries, it attaches utmost importance to the cooperation among developing countries as a means to strengthening their science and technology capacities ("collective self-reliance"). Moreover, it underscores the potential of the United Nations system and other international organizations to contribute to achieving the programme objective in this first target area.

Under the second target area, the Vienna Programme of Action covers the bilateral and multilateral scientific-technological relations among states in general and between industrially developed and developing countries in particular. With respect to possible regulatory systems affecting the transfer of technology and the activities of transnational corporations, which have been under consideration in various United Nations organs and have even become—in the case of the transfer of technology—the subject of formal diplomatic negotiations, the Programme of Action was substantially curtailed as compared with the draft of the Group of 77. It reaffirms the need for regulation in general terms, but it does not provide new, specific directions for the ongoing negotiation processes.

Conversely, this part of the Programme of Action contains fairly detailed recommendations for international cooperation in the field of scientific and technological information. This cooperation is to have at its peak a global information network run by the United Nations that is to facilitate access to national information systems. However, information of a proprietary nature, particularly information that qualifies as company secrets in industrialized countries, is not covered by these recommendations.

Finally, the Vienna Programme recommends protection and control provisions, some of which are quite far-reaching, relating to international research cooperation for the benefit of the developing countries. The general aim of these provisions is to transform the status of developing countries from being primarily passive to becoming more active participants on an equal footing in research projects that have a direct bearing on their territories and societies.

The third target area of the Vienna Programme of Action comprises recommendations for institutional innovations in United Nations policymaking and the transfer of resources with regard to science and technology for development.[61] The first element in this set of institutional innovations is the setting up of a standing Intergovernmental Committee on Science and Technology for Development (ICSTD) that is open to the participation of all states and is to replace the fifty-four-member Committee on Science and Technology for Development of the Economic and Social Council. This new Intergovernmental Committee should be authorized to report to the General Assembly through the Economic and Social Council and to work out proposals for the implementation of the Vienna Programme as well as guidelines for coordinating the activities of all member units of the United Nations system in the field of science and technology for development.

A second institutional innovation is envisaged by the recommendation to establish an independent and high-ranking unit for science and technology for development within the United Nations secretariat that would replace the Office for Science and Technology of the Department of International Economic and Social Affairs. The Vienna Programme also recommends that this new secretariat unit assist the Director-General for Development and International Economic Cooperation[62] in his task of providing overall guidance and coordination within the United Nations system. In addition, the Director-General himself is called upon to render assistance to a new committee of governmental experts that would be entrusted with elaborating detailed proposals for a United Nations financing system for science and technology for development.

The third institutional innovation recommended by the Vienna Programme deals with new mechanisms of mobilizing external financial resources for the promotion of science and technology in the developing countries. More specifically, the Programme of Action recommends, as a short-term arrangement, the creation of an

Interim Fund for the years 1980 and 1981 to be financed by voluntary contributions in the amount of at least US$250 million and to be administered by the United Nations Development Programme. The guidelines for the allocation of the Fund's resources are to be laid down by the General Assembly and the Intergovernmental Committee, whereas the operational activities of the Interim Fund would be overseen by the Governing Body of UNDP. Over and above this short-term arrangement, the Vienna Programme mandates the establishment of long-term financial arrangements for a United Nations financing system for science and technology for development that would take into account, *inter alia*, "the need for predictability and continuous flow of financial resources" as well as "the need for substantial resources in addition to those that now exist within the United Nations system" (paragraph 112). However, the shape of these long-term arrangements is to be studied by the already mentioned committee of governmental experts, which should report on its findings to the General Assembly at its thirty-sixth session, i.e., in late 1981.

B. Obstacles to Consensus Formation

The approval of the Vienna Programme of Action—even in the truncated form in which it was finally adopted by UNCSTD—was preceded by a very laborious process of consensus formation. The difficulties faced by UNCSTD in achieving consensus on a programme of action are best indicated by the fact that more than thirty paragraphs in the draft programme of action submitted by the Group of 77 did not survive the negotiating process. These paragraphs, deleted from the Programme of Action as adopted, appear, in some instances modified, in two annexes to the Vienna Programme of Action.[63] What saved UNCSTD from failing to reach agreement on a programme of action was the basic willingness to compromise and a considerable flexibility in the actual negotiations on the part of all Conference participants. The most far-reaching concessions—as compared to their initial positions—were undoubtedly made by the Group of 77, i.e., the developing countries. But the industrially developed countries, too, abandoned their stands on a number of points, particularly with respect to institutional changes under the third target area. This meant that UNCSTD did lead to an agreement on at least an embryonic form of a planning, regulatory, and financing system for science and technology within the framework of the United Nations system.

Nevertheless, the initial probability of nonagreement or, conversely, the efforts needed for engineering consensus on a programme of action for science and technology for development to be adopted by UNCSTD should not be underestimated. First of all, a characteristic feature of consensus formation within the United Nations system is the necessarily large number of participants in any process of consensus formation: There are more than 150 Member States, of which more than 120 took part in UNCSTD, and there are more than thirty relatively independent institutions within the United Nations system, of which, again, a good dozen were directly involved in and affected by UNCSTD. Whenever a large and diverse number of actors are in a position to have, or to claim, a stake in deliberations and negotiations on relatively complex issues, the process of consensus formation is likely to be stalled unless formal and informal devices to reduce the number of actual participants are used with both circumspection and determination.

The difficulties of mobilizing consensus at world conferences of the United Nations system are further increased by linkage strategies, i.e., attempts by participants to influence, or even override, ongoing processes of negotiations over related, or even unrelated, issues in other international fora. For example, at the UNCSTD, the Group of 77 sought to strengthen its negotiating position, through the programme of action to be adopted by UNCSTD, in the conferences on a code of conduct for the transfer of technology, a code of conduct for transnational corporations, and the revision of the Paris Convention on the Protection of Industrial Property.

More decisively, however, the probability of agreement or nonagreement depends on the prior divergence or convergence of substantive interests of the conference participants. On the one hand, there is a genuine conflict of interest between developing and industrially developed countries rooted in the structure of an international division of labour which, at least so far, has not contributed to narrowing the wide gap in scientific and technological capacities between them.[64] This conflict of interest also exists in the perceptions of those most concerned: The representatives of developing countries have seized upon the opportunity afforded by a United Nations conference on the subject to express their dissatisfaction with the scientific and technological dependence of their countries and to voice proposals for overcoming this dependence that are, more often than not, at variance with the preferences of industrially developed countries. On the other hand, the developed countries, and the representatives of many Western industrial states in particular, exhibit concern about the consequences of a possible decline of their comparative advantages on world markets (and in power politics) that may result from failing to maintain their scientific and technological superiority. Yet what most Western industrial countries consider a fair mechanism for achieving a balance of interest between themselves and the developing countries, the commercial transfer of technology, has been the subject of much contention.[65] However, although the developing countries have never roundly rejected it, they have come to insist on the necessity of regulating it through international and national public policies (including new legal norms). Conversely, the industrially developed countries do not categorically deny the desirability of setting up an international regulatory regime for the transfer of technology and do not oppose all practical efforts to do so, although they continue to have reservations about the scope and intensity of such regulation.

A greater convergence of interests between developing and industrially developed countries is evident in other areas. For industrial countries, which dominate world markets, and for the transnational corporations domiciled in them, the strengthening of the scientific and technological capacities of the developing countries does not, at least in principle, constitute a threat to their international economic and political standing. Therefore, the provision of financial resources for this purpose is met with resistance less for strategic than for fiscal reasons. Indeed, the emergence of a stronger and more reliable scientific and technological infrastructure in developing countries would consolidate the foundations for the smooth further development of international economic interdependence. However, the apparent reluctance on the part of some Western industrial countries in particular to act upon the accepted goal of building up the scientific and technological bases of developing countries is born out of fear that developing countries might use their newly acquired capacities for politically motivated discrimination among supplier countries and corporations. Whether or not this fear is real and/or expressed for propagandistic purposes, most developing countries have not built up a record of discriminatory policies toward the industrial countries. Rather, they have tended to pursue accommodative policies, not the least important reason being not to disturb the inflow of external financial resources, which have provided many Third World regimes with a margin of economic and political flexibility.

Although we can therefore proceed from a mixed bag of converging and diverging interests in the relationships between the Western industrial countries and the developing countries, this distribution of interests does not exhibit a significantly simpler structure as far as the relationships between the Eastern industrial countries and the Third World are concerned. However, these relationships have been, at least until recently, much less controversial, as the relevant transaction flows have been smaller by far. The Eastern industrial countries insist that the principle of nondiscrimination be observed across the whole range of international transactions. In their international practice Eastern industrial countries usually attempt to achieve acceptance of strict reciprocity (mostly of a bilateral nature). In general, they tend to support the developing countries in their claims for a redistribution of the world's scientific and technological potential; but they leave no doubt either that they are neither willing nor in a position to assume any additional responsibility for it. Moreover, following their general line of resisting institutional task expansion within the United Nations system, the Eastern industrial countries take a skeptical, if not a negative, view of endeavours to set up a planning, regulatory, and financing system for science and technology within the United Nations system. This general line notwithstanding, the Eastern industrial countries have been flexible enough not to prevent the formal registration of a consensus in these matters once a compromise had been reached between Western industrial countries and developing countries.

Any retrospective assessment of the initial probability of nonagreement on substantive agenda items of UNCSTD must be based on relatively rough indications derived from participant observation,[66] which, in research of this kind, operates as a substitute for more refined methodological instruments. The findings as to the number of necessary participants, the linkage strategies pursued by the Group of 77, and the configuration of interests among the major groups of states taking part in the Conference lead to the conclusion that consensus at UNCSTD could not be expected to be achieved easily. Indeed, the analysis points up considerable, though not insurmountable, obstacles to substantive agreements on a number of issues before the Conference. How the agreements on intensified policy coordination in the field of science and technology for development were brought about and how certain obstacles were overcome will be examined below.

C. *Strategies of Programmatic Consensus Formation*

In the course of Conference preparations and during UNCSTD itself, the major actors taking part in the Conference had recourse at various times to different strategies in order to find a common denominator for a programme of action in general and for the strengthening of policy coordination in the field of science and technology for development through the United Nations system in particular. On the whole, although efforts to upgrade the common interest or to compromise on the basis of splitting the difference were clearly present, adaptive strategies of conflict avoidance, i.e., of reducing the need for consensus and settling on the basis of the minimum common denominator, prevailed.[67] This became ever more obvious as the preparatory process drew to a close and throughout the Conference itself. Moreover, such strategies were preferred by the major Western industrial countries almost from the beginning of the preparatory period, after they had realized that they could not prevent this Conference from becoming part and parcel of the more general North-South controversy about the establishment of the New International Economic Order.

In contrast, the developing countries hoped to gain acceptance for a far-reaching programme of action that was to aim at solving the major problems of developing countries in the field of science and technology, i.e., the lack of a "critical mass" of scientific and technological activities within most developing countries and their inequitable share of the overall resources for, and benefits derived from, science and technology. To achieve this goal, the Group of 77 emphasized information and persuasion strategies vis-à-vis the developed countries. Its efforts to establish policy legitimacy[68] for the measures to be included in the Programme of Action was to have its cognitive foundation in the national papers and its normative foundation in the General Assembly resolutions ushering in the Second Development Decade and calling for the establishment of the New International Economic Order. However, as the Conference neared and the negotiations over the draft programme of action remained deadlocked on a number of crucial elements, such as an international regime for the transfer of technology, the provision of new financial resources for science and technology in the Third World, and the restructuring of United Nations institutions dealing with science and technology, the pursuit of information and persuasion strategies was replaced, reluctantly and slowly, by a readiness on the part of the Group of 77 to shelve controversial parts of the draft programme of action and to defer putting them to a vote.

A decisive factor in reaching agreement on the Vienna Programme of Action at all was the de facto recognition of the consensus rule by all states participating in the conference, i.e., the waiver of the right to insist on majority decisions. This recognition of the consensus rule, particularly by the developing countries, reflected the appreciation of the fact that the implementation of a programme of action would be gravely endangered if the industrially developed countries remained passive from the outset. Conversely, the consensus rule put pressure even on the most status quo–oriented Western industrial countries not to appear intransigent and not to withdraw from United Nations fora. The compulsion to compromise with the industrially developed countries and often to settle on terms closer to their original positions than vice versa, which the developing countries thus imposed upon themselves, was, and still is, not undisputed in their own ranks. At UNCSTD, too, at least occasionally the question was raised by Third World delegates whether the compromises that could be obtained at this Conference were worth the price of abstaining from decisions on policy concepts and action recommendations through majority vote.

On comparing UNCSTD with other ad hoc world conferences sponsored by the United Nations system, we can see that its results have been no less significant than those of most other conferences and its capacity for consensus formation did not fall below average. Certainly, a more far-reaching consensus could have been generated if the decision situation in which the Conference participants found themselves had been affected by an acute international crisis in North-South relations, or if the issue area dealt with at the Conference had attracted worldwide public attention for other reasons. Unlike some of the global conferences of the early 1970s, such as the United Nations Conference on the Human Environment[69] in 1972 and the United Nations World Population Conference[70] and the World Food Conference,[71] both in 1974, UNCSTD did not attract input from the political environment that would facilitate agreement on major policy innovations. One reason for the comparatively low political profile of UNCSTD derives from the fact that scientific and technological dependence, the substantive focus of UNCSTD, cannot be made nearly as visible as manifestations of environmental destruction, overpopulation, or starvation. Thus, at UNCSTD, a relatively spontaneous increase in the level of demands for strengthening internationally organized problem-solving activities did not materialize.

An examination of the consideration given to various interests in the Vienna Programme of Action must start out at the level of government-mediated representation of interests. As already indicated above, the temporizing resistance of the Western industrial countries (with the exception, in some instances, of the Nordic countries) during the consultations and negotiations about a draft programme of action paid off insofar as they had to agree to only comparatively slight concessions concerning the setting up, within the United Nations, of a planning, regulatory, and financing system for science and technology for development that was to have but modest powers and resources. The representatives of most of the Western industrial countries felt themselves constrained to prevent the negotiating process from leading to results that might have been interpreted, in their various domestic political systems, as encroachments, even though marginal, on private property rights and as the initiation of a global fiscal adjustment system operating on the basis of more or less automatic transfers of financial resources. The Western industrial countries' approach to negotiations throughout the conference process made it clear that the governments and the hegemonic social groups in these countries shared the view that structural adjustments for the benefit of developing countries should not be allowed to become the subject of international policymaking over which they do not exercise a high degree of control.

The extent of international policy-coordination in the field of science and technology for development that was agreed upon at UNCSTD does not imply that international political interventions to apply brakes to customary transaction flows in science and technology would become possible, even if such flows reinforced the dependence of developing countries. Thus, de facto, the Conference outcome provides a mirror image of the international stratification in science and technology. For the private and public owners and suppliers of technologies in industrial countries and in some cases also in developing countries, UNCSTD did no harm, as political controls of exchange processes were fought off. For those developing countries that import advanced technologies and possess bargaining leverage because of their resource endowments and/or market opportunities, the provision of additional financial means, even of modest volume, is welcome: Additional scientific and technological infrastructure projects may enhance their attractiveness to foreign investment. Finally, the Vienna Programme contains little to satisfy the interests of the great majority of developing countries, which are truly underdeveloped, if not undeveloped, with respect to science and technology. For them, the Vienna Programme of Action offers but the all too familiar "raindrop in the desert," at least in the short term.

The few tangible results of UNCSTD, irrespective of their differential objective value for various groups of developing countries, could hardly have been obtained without the unbroken, though at times precarious, solidarity of the Group of 77. This solidarity can be explained, first of all, by its overall strategic value for the developing countries in dealing with the industrially developed countries, which transcends individual North-South encounters. Second, the frequently noticeable intransigence of the major Western industrial countries in particular leaves the developing countries with no other alternative than to stick together and agree on a common platform, even if this means simply drawing up a list of demands without any attempt at integrating them. Finally, the joint action compensates for their individual political weaknesses – though at the price of conceding leadership powers by one segment of the Group of 77 (those developing countries that cannot afford to play a strong international role) to another. At UNCSTD, the dominant role that devolved upon some Latin American states as well as countries such as India and Tunisia is but an illustration of this process.

6. Programme Implementation and Prospects of Policy Coordination After UNCSTD

A. Review and Approval of the Vienna Programme of Action by the General Assembly

The Vienna Programme of Action adopted by consensus at UNCSTD had formally the status of a recommendation to the United Nations General Assembly. During its thirty-fourth session, the Second Committee dealt with the Programme of Action in great detail. The comprehensive resolution that the General Assembly adopted to approve of the outcome of UNCSTD represented a first step on the way to putting the Programme of Action into effect.[72] In so doing the General Assembly found it necessary to find a more precise formulation for some provisions of the Programme of Action, to supplement it with more detailed guidelines for action, and even to expand on it. The Negotiating Groups of the Second Committee that were commissioned to draft a resolution on UNCSTD experienced hard bargaining sessions, particularly when it appeared that the Group of 77 might attempt to reopen issues that, in the view of most industrially developed countries, had been settled in Vienna. At times, the negotiations seemed to be on the verge of replaying the drama of UNCSTD. However, in the end the notion prevailed that the Negotiating Groups should stick to the Vienna texts as closely as possible.

A more concrete elaboration of the Vienna Programme's recommendations proved necessary with regard to the new unit for science and technology within the United Nations secretariat. To facilitate consensus in Vienna the provisions of the Programme of Action relating to a new secretariat structure had to remain somewhat vague;[73] to avoid a head-on collision it was decided to request "the Director-General, under the authority of the Secretary-General" to submit a report to the General Assembly to facilitate a final decision on three aspects of this issue: the rank of the new secretariat head (i.e., Assistant or Under Secretary-General), the location of the new unit within the secretariat's overall hierarchical structure (i.e., as part of the Department of International Economic and Social Affairs or as semi-independent entity reporting directly to the Director-General for Development and International Economic Cooperation), and the future of the existing Office for Science and Technology and of its staff in particular.[74] After protracted negotiations, the Second Committee could not reach a consensus on these questions. The member states of the European Community as well as the Soviet Union and East European countries had strong reservations about assigning too high a rank to the new secretariat unit and moving it out of the Department of International Economic and Social Affairs; they also were clearly unhappy about the prospect of losing senior executive post(s) to the developing countries. The solution on which the Group of 77 and the remaining industrial countries, notably the United States and the Nordic countries, were able to agree provided for the creation of a Centre for Science and Technology for Development, "as a new, organizationally distinct entity," headed by an Executive Director with the rank of Assistant Secretary-General "who shall be responsible to and report directly to the Director-General for Development and International Economic Cooperation." It was further decided to dissolve the Office for Science and Technology and to allocate its established posts and other budgetary resources to the new Centre. However, the Group of 77 met with strong resistance on the part of the industrially developed countries to its demands for additional staff posts in the new Centre.[75]

The Vienna Programme of Action deliberately left it to the General Assembly to supplement the programme by working out provisions for the operation of the envisaged Interim Fund: Guiding principles, procedures, and organization of the Interim Fund required detailed definitions and regulations, and this provided one of the already mentioned opportunities for renewed tense bargaining between the Group of 77 and the chief donor countries. Initially, i.e., before the Second Committee took up the agenda item on UNCSTD, it appeared that the so-called Prospectus for the Interim Fund had been successfully prenegotiated in an informal process involving representatives of the Administrator of UNDP, the Director-General for Development and International Economic Cooperation, and both the chief donor countries and the Group of 77.[76] The main function of the Prospectus was to translate into technical provisions the delicate balance between the policymaking competence of the newly established Intergovernmental Committee and the executive competence of the UNDP Governing Body. This could not be expected to be an easy task, as it meant reconciling the interest of the developing countries in strengthening the Intergovernmental Committee with the interest of the chief donor countries in preserving the jurisdiction of the UNDP Governing Body over project approval and the supervision of project execution. When the Group of 77 did submit an alternative prospectus for the Interim Fund[77] that would have shifted this balance to the advantage of the Intergovernmental Committee, the chief donor countries rose in massive protest and demanded its withdrawal (with the implied threat of withholding contributions to the Interim Fund). In the end, the Group of 77 did not insist on treating its alternative prospectus as a basis for negotiations and contented itself with a number of amendments to the draft prospectus submitted by the administrator of UNDP, which were acceptable to all Member States present.[78]

The General Assembly resolution on UNCSTD went beyond the provisions of the Vienna Programme of Action by calling, on a motion of the Canadian delegation, for a comprehensive, critical stocktaking of the activities of the organizations of the United Nations system in the field of science and technology for development and for assessing their efficiency.[79] Although this measure is intended to improve the coordination of science and technology policies and activities in the United Nations system, it will have to reckon with the widespread bureaucratic egocentricity of various secretariats belonging to the United Nations system. It might be argued that, unless the defective or even nonexistent coordination of policies on the part of many Member States toward, and within, the intergovernmental organs of the United Nations system is overcome, efforts at improving coordination at the level of the United

Nations system alone will prove to be largely futile.[80] The segmented, sometimes clientilistic, relationships between international secretariats in the United Nations system and Member States' delegates representing the corresponding policymaking sector at the national level constitute a formidable stumbling block on the road to effective coordination of sectoral policies and activities at the level of the United Nations system.

B. A New Institutional Framework for the Implementation[81] of the Vienna Programme of Action

With the first session of the new Intergovernmental Committee on Science and Technology for Development, 28 January–1 February 1980 in New York,[82] the phase of implementing the Vienna Programme of Action as mandated by the Thirty-Fourth General Assembly commenced, also in a formal sense. The question that interests us here is to determine what practical steps, if any, have been taken by governments and international secretariats to make progress toward achieving one of the central objectives of UNC-STD—the setting up of a planning, regulatory, and financing system for science and technology for development within the United Nations system. While it is still too early to come up with valid assessments of whether or not this objective will be achieved, it seems permissible to conjecture about trends.

The setting of an early date for the first session of the Intergovernmental Committee notwithstanding, the process of implementing the Vienna Programme had a slow start and has not yet gained a strong momentum. The new secretariat unit did not really become operational until mid-1980, the Committee of Governmental Experts for a United Nations financing system for science and technology for development did not get down to substantive work before the end of 1980, and the provision of expert advice to the Intergovernmental Committee took even longer to be organized. These are, after all, elementary components of the institutional infrastructure for the implementation of the Vienna Programme. The slow beginnings of the implementation process can surely be attributed to aftereffects of controversies during UNCSTD and the Thirty-Fourth General Assembly. Although the lingering on of earlier controversies need not be overestimated, they will probably cause delays in schedules for implementing a number of intermediate steps provided for by the Vienna Programme, such as the Operational Plan for implementing the Vienna Programme or the report to be submitted by the Committee of Governmental Experts.

More serious obstacles to implementing the Vienna Programme of Action arose regarding the mobilization of new financial resources to put into effect the new Interim Fund for Science and Technology for Development. Reaching agreement on additional guidelines for project selection by the Interim Fund did not pose serious problems during the second session of the Intergovernmental Committee (22 May–4 June 1980),[83] but the level of funding has fallen short of the target by far. At the first pledging conference for the Interim Fund (27 March 1980), firm pledges and contributions amounting only to US$35.8 million were announced by thirty-five countries.[84] Despite this disappointing result the Administrator of UNDP declared the Interim Fund operational on 19 May 1980—a bold step designed to influence reluctant donor countries to adopt a more forthcoming attitude toward the Interim Fund. In addition, the small working staff that the Administrator of UNDP had assembled in 1979 was quickly enlarged to proceed rapidly and in direct consultations with interested developing countries with the identification of suitable projects and with the processing of project applications already submitted by them.

There is apparently no lack of projects that would come under the purview of the Interim Fund's mandate, and the available evidence indicates that the identifiable needs of developing countries for funding such projects go well beyond the target of US$250 million set for the Interim Fund. Nonetheless, subsequent pledges and contributions increased resources available to the Interim Fund to not more than about US$50 million, largely because chief donor countries, such as the United States and the Federal Republic of Germany, have so far refrained from making substantial and firm commitments to the Interim Fund. This has been especially disturbing in the case of the United States, because its delegation belonged to the main architects of the Vienna compromise package of which the Interim Fund represented a crucial ingredient. In view of the reluctance or lack of commitment on the part of traditional donor countries, efforts have been made by some developing countries together with representatives of The Interim Fund to attract funds from other sources, above all from member states of the Organization of Petroleum Exporting Countries with large current-account balance of payments surpluses. There are no clear indications yet whether or not such an initiative is likely to succeed. Taken together, these developments do not augur well for the politically even more sensitive task of the Committee of Governmental Experts to come up with agreed upon, innovative recommendations for long-term financial arrangements for science and technology development.

7. Conclusion

UNCSTD has been a significant episode in the evolving North-South dialogue.[85] It certainly did not break new ground nor did it reverse this process. The Conference has clarified items on the agenda of the North-South dialogue, but it did not necessarily lighten the burden of those participating in it. A global compact (or "Pact for Interdependent Growth")[86] between developing and industrially developed countries has not been advanced very far by UNCSTD, certainly far less than one would have hoped. Yet, the very shortcomings of the Conference process as well as of the implementation of its results point to both the deep crisis in the North-South dialogue and the urgency of redoubling efforts to strengthen international policy-coordination for development. The contemporary dislocation of the world economy and its consequences for both the industrially developed countries and most of the developing countries tend to favour shortsighted policies of retrenchment on one side and of radicalization or "muddling-through" on the other side. Fatalistic resignation in both industrially developed and developing countries represents the greatest danger of all.

Notes

1. *Report of the United Nations Conference on Science and Technology for Development*, Vienna, 20–31 August 1979 (New York: United Nations, 1979), pp. 8–9 (A/CONF.81/16; Sales No.: E.79.I.21).

2. Ibid., p. 14.

3. Ibid.

4. Ibid., pp. 15–16.

5. Summary of the Introductory Statement by the Secretary-General of the Conference on Item 70 ("United Nations Conference on Science and Technology for Development") to the Second Committee of the General Assembly, 16 November 1979, p. 1 (mimeo).

6. Klaus-Heinrich Standke, "The Prospects and Retrospects of the United Nations Conference on Science and Technology for Development," in *Technology in Society*, Vol. 1 (1979), pp. 353–386, especially p. 383.

7. Norman Graham and Stephan Haggard, "Diplomacy in Global Conferences," *UNITAR News*, Vol. 11 (1979), pp. 14–21, especially pp. 14–15.

8. I am attempting here to draw on a paradigm for the study of policy coordination and interorganizational decision making in federal states developed by Fritz W. Scharpf and others. See, in particular, Fritz W.

Scharpf, Bernd Reissert, and Fritz Schnabel, *Politikverflechtung* (Kronberg: Scriptor, 1976), especially Chapter 1 by Fritz W. Scharpf, to which I refer frequently hereafter. Another book bearing on this general subject is Kenneth Hanf and Fritz W. Scharpf (eds.), *Interorganizational Policy Making* (London and Beverly Hills: Sage, 1978).

9. Cf. on this point Joachim Betz, *Die Internationalisierung der Entwicklungshilfe* (Baden-Baden: Nomos, 1978); and by the same author, "The Internationalization of Development Policy," *Intereconomics*, Vol. 14, No. 1 (1979), pp. 25–31.

10. Even as ardent a believer in world government as Saul Mendlovitz, who wrote in 1974 that "there is no longer a question of whether or not there will be world government by the year 2000," had to concede six years later that it is unlikely to be that close. See his "General Introduction" to Johan Galtung, *The True Worlds* (New York: Free Press, 1980), pp. XIII–XXII, especially pp. XXI–XXII.

11. This distinction between two types of international policy-coordination is derived, after appropriate adaptation, from Fritz W. Scharpf, op. cit. (note 8), pp. 34–35.

12. Cf., for example, Philippe C. Schmitter and Gerhard Lehmbruch (eds.), *Trends Toward Corporatist Intermediation* (London and Beverly Hills: Sage, 1980).

13. See note 8.

14. It should be noted that, in this regard, I have shifted the preference I showed in earlier studies on this subject; however, the empirical analysis contained in these studies has not been invalidated as a consequence. Cf. Volker Rittberger, *The New International Order and United Nations Conference Politics: Science and Technology as an Issue Area*, Science and Technology Working Papers Series No. 1 (New York: UNITAR, 1978); "United Nations Conference Politics and the New International Order in the Field of Science and Technology," *Journal of International Affairs*, Vol. 33, No. 1 (1979), pp. 63–76.

15. In this context, "joint international task" refers to a goal- or problem-oriented politico-administrative action pattern that may be described as a planning, regulatory, and financing system, however rudimentary, involving relevant nation-state units and multinational policymaking institutions. We can speak of a planning, regulatory, and financing system when there are joint institutional capacities for information gathering and processing and for programme formulation and implementation and a modicum of joint resources mobilization to fulfill the joint task(s).

16. Brigitte Schroeder-Gudehus, "Science, Technology and Foreign Policy," in Ina Spiegel-Rösing and Derek de Solla Price (eds.), *Science, Technology and Society* (London and Beverly Hills: Sage, 1977), pp. 475–476. Cf. also M. P. Crosland, "The History of the International Organisation of Science: Some Preliminary Sketches," in Frank R. Pfetsch (ed.), *Internationale Dimensionen der Wissenschaft* (Erlangen: Institut für Gesellschaft und Wissenschaft, 1979), pp. 37–60.

17. A detailed factual overview is given by Jean Touscoz, *La Coopération Scientifique Internationale* (Paris: Editions Techniques et Economiques, 1973).

18. General Assembly Resolution 1710 (XVI), in General Assembly Official Records (*GAOR*), *16th Session, Supplement No. 17* (A/5100), pp. 17–18.

19. General Assembly Resolution 2626 (XXV), in *GAOR, 25th Session, Supplement No. 28* (A/8028), pp. 39–49.

20. General Assembly Resolutions 3201 and 3202 (S-VI), in *GAOR, 6th Special Session, Supplement No. 1* (A/9559), pp. 3–12; General Assembly Resolution 3362 (S-VII), in *GAOR, 7th Special Session, Supplement No. 1* (A/10301), pp. 3–9. Equally important in this context is the Charter of Economic Rights and Duties of States proclaimed by General Assembly Resolution 3281 (XXIX), in *GAOR, 29th Session, Supplement No. 31* (A/9631), pp. 50–55.

21. General Assembly Resolution S-11/23, in *Resolutions and Decisions of the General Assembly at its Eleventh Special Session, 25 August–15 September 1980,* in GAOR, 11th Special Session, Supplement No. 3 (A/S-11/3), p. 7.

22. See note 18.

23. *Science and Technology for Development. Report on the United Nations Conference on the Application of Science and Technology for the Benefit of the Less Developed Areas.* 8 vols. (New York: United Nations, 1963) (E/CONF.39/1; Sales No.: E.63.I.28). A brief assessment of this conference can be found in Klaus-Heinrich Standke, "Wissenschaft und Technologie im System der Vereinten Nationen," *Vereinte Nationen,* Vol 24, No. 1 (1976), p. 10.

24. On the involvement of ACAST in the preparations for the Second Development Decade see Office for Science and Technology, *Science, Technology and Global Problems. The United Nations Advisory Committee on the Application of Science and Technology for Development* (New York: Pergamon, 1979), especially pp. 4–5, 35–36.

25. Cf. paragraphs 60–64 of General Assembly Resolution 2626 (XXV).

26. Resolution 1621 (LI) of the Economic and Social Council, in *Economic and Social Council Official Records, 51st Session, Supplement No. 1* (E/5073), 1971, pp. 21–22.

27. Cf. section III, paragraph 7 of General Assembly Resolution 3362 (S-VII). The report of the Intergovernmental Working Group of the CSTD is available as United Nations document E/C.8/28 of 6 August 1975.

28. Resolutions 2028 (LXI) and 2035 (LXI) of the Economic and Social Council, in *Economic and Social Council Official Records, 61st Session, Supplement No. 1* (E/5889), pp. 10–11 and 15–16. The substantive items of the conference agenda were already formulated in ECOSOC resolution 2028 (LXI), and they did not change afterwards, except for the last. They are:

1. Science and technology for development:

    a. The choice and transfer of technology for development;

    b. Elimination of obstacles to the better utilization of knowledge and capabilities in science and technology for the development of all countries, particularly for their use in developing countries;

    c. Methods of integrating science and technology in economic and social development;

    d. New science and technology for overcoming obstacles to development.

2. Institutional arrangements and new forms of international co-operation in the application of science and technology:

    a. The building up and expansion of institutional systems in developing countries for science and technology;

    b. Research and development in the industrialized countries in regard to problems of importance to developing countries;

    c. Mechanisms for the exchange of scientific and technological information and experience significant to development;

    d. The strengthening of international co-operation among all countries and the design of concrete new forms of international co-operation in the fields of science and technology for development;

    e. The promotion of co-operation among developing countries and the role of industrially developed countries in such co-operation.

3. Utilization of the existing United Nations system and other international organizations:

Utilization of the existing United Nations system and other international organizations to implement the objectives set out above in a co-ordinated and integrated manner.

4. Science and technology for the future:

Debate on the basis of the report of a panel of experts to be convened on this subject.

For the Conference agenda as finally adopted see *UNCSTD Report*, op. cit. (note 1), p. 11.

29. General Assembly Resolution 31/184, in *GAOR, 31st Session, Supplement No. 39*, pp. 86–87.

30. An informative description of this early phase of the preparations for UNCSTD is given in the unpublished Ph.D. dissertation of Mary M. Lyon-Allen, "The United Nations Conference on Science and Technology for Development: The International Negotiation of Technological Relations," Washington, D.C., George Washington University, 1979.

31. Economic and Social Council Resolution 2028 (LXI), paragraph 3 (see note 28).

32. It can be said that the agenda of the Conference and the basic features of the preparatory period were already decided upon in 1975 and 1976, i.e.,

during the meeting of the Intergovernmental Working Group of the CSTD
(see note 27), the third session of the Committee on Science and Technology
for Development, 2–20 February 1976 (see *Economic and Social Council Official Records, 61st Session, Supplement No. 3* [E/5777; E/C.8/43], pp.
17–29), and the 61st session of the Economic and Social Council in the summer of the same year (see note 28).

33. *Report of the Preparatory Committee for the United Nations Conference on Science and Technology for Development,* in *GAOR, 32nd Session, Supplement No. 43* (A/32/34).

34. Ambassador da Costa had been intimately involved in the process of
deciding on the holding of UNCSTD, having been the chairman of the Intergovernmental Working Group and of the CSTD at its second and third
sessions.

35. The "wise men" approach to the preparation of conference documentation was strongly felt at the Stockholm Environment Conference (1972),
for instance. At the Rome Food Conference (1974) the documentation derived to a considerable extent from background material supplied by the
United States. The emphasis, during the preparations for UNCSTD, on national papers from as large a cross-section of Member States as possible to
serve as the principal underpinning of Conference documentation seems
quite reasonable from the point of view of the developing countries,
especially the more advanced of them.

36. The guidelines are contained in Decision 2(I) of the Preparatory Committee. *Report of the Preparatory Committee for UNCSTD,* op. cit. (note
33), pp. 25 *et seq.*

37. Ibid., p. 6.

38. The list of five subject areas is contained in Resolution 3(II) of the
Preparatory Committee. It runs as follows:

1. Food and agriculture:
   a. Agriculture technology and techniques and their improve-
      ment;
   b. Nutrition;
   c. Fisheries;
   d. Food storage and processing.
2. Natural resources including energy:
   a. Renewable and non-renewable;
   b. Conventional and non-conventional sources of energy;
   c. Development and conservation;
   d. Rational management and utilization.
3. Health, human settlement and environment:
   a. Medicinal plants and pharmaceuticals;
   b. Health services;
   c. Housing;
   d. Social services and environment.
4. Transport and communications.
5. Industrialization, including production of capital goods. Cf.

Report of the Preparatory Committee for the United Nations Conference on Science and Technology for Development, in GAOR, 33rd Session, Supplement No. 43 (A/33/43), pp. 22–23.

The controversy about the subject areas can even be gleaned from the official report on the first session of the Preparatory Committee; cf. *Report of the Preparatory Committee for UNCSTD*, op. cit. (note 33), p. 6.

39. On the International Scientific Colloquium sponsored by ACAST see Klaus-Heinrich Standke and M. Anandakrishnan (eds.), *Science, Technology and Society: Needs, Challenges and Limitations* (New York: Pergamon, 1980).

40. Decision 1(I), paragraphs 5–7, in *Report of the Preparatory Committee for UNCSTD*, op. cit. (note 33), pp. 21–22.

41. Decision 4(II), in *Report of the Preparatory Committee for UNCSTD*, op. cit. (note 38), pp. 27–33, especially p. 30.

42. Decision 1(I), paragraphs 8–9, in *Report of the Preparatory Committee for UNCSTD*, op. cit. (note 33), p. 22.

43. Decision 20(IV), in *Report of the Preparatory Committee for UNCSTD*, in *GAOR, 34th Session, Supplement No. 43*, Vol. II (A/34/43, Vol. II), p. 15, and Decision 25(V) in *Report of the Preparatory Committee for UNCSTD*, in *GAOR, 34th Session, Supplement No. 43*, Vol. III (A/34/43, Vol. III), p. 15.

44. Cf. Standke, op. cit. (note 6), pp. 362–363. This advice, if it was not meant to be cynical, must assume the existence of a kind of political process within Member States that is open to the participation of nongovernmental groups; yet it is safe to state that the facts do not bear out this assumption in its generalized version

45. The national papers are part of the official conference documentation and can be located under the symbol A/CONF. 81/NP 1 et seq.

46. See Jurg Mahner, *A Preliminary Assessment of National Papers as Basis for UNCSTD Conference Programming*, Science and Technology Working Papers Series, No. 3 (New York: UNITAR, 1979), p. 17.

47. United Nations document A/33/303, "United Nations Conference on Science and Technology for Development. Draft Outline of the Programme of Action. Note by the Secretary-General."

48. General Assembly Resolution 33/192, paragraph 2(b), in *GAOR, 33rd Session, Supplement No. 45* (A/33/45), pp. 120–121.

49. United Nations document A/CONF.81/PC/21, "United Nations Conference on Science and Technology for Development. Preliminary Draft Programme of Action. Note by the Secretary-General of the Conference."

50. Resolution 6 (III), in Report of Preparatory Committee for the United Nations Conference on Science and Technology for Development, in *GAOR, 33rd Session, Supplement No. 43*, Vol. I (A/33/43, vol. I), p. 15.

51. United Nations document A/CONF.81/PC/28, "United Nations Conference on Science and Technology for Development. Preliminary Draft Programme of Action (rearranged as required by the Preparatory Committee

in Resolution 6 [III]). Note by the Secretary-General of the Conference."

52. Decision 10 (III), in *Report of the Preparatory Committee for UNC-STD*, op. cit. (note 50), p. 17.

53. On the work of the Preparatory Committee during these sessions and the documents before it see the reports referred to in note 43 above.

54. For a more detailed description and discussion of these approaches and concepts see Chapter 1, section 5.

55. See *Report of UNCSTD*, op. cit. (note 1), passim.

56. This point has been stressed by Lothar Brock, *Entwicklungsnationalismus und Kompradorenpolitik* (Meisenheim: Anton Hain, 1975).

57. See, for example, *Report of UNCSTD*, op. cit. (note 1), p. 19 and pp. 111 *et seq.* ("Science and Technology and the Future").

58. Ibid., pp. 23, 48, 42–44 (Resolution on Women, Science and Technology).

59. Cf., for instance, the Conference document, *Science and Technology and the Concept of Development*. Consolidated Discussion Paper Prepared by the Secretary-General of the Conference (A/CONF.81/4) as well as the supplementary document representing the (lowest) common denominator for the vested interests of the various United Nations bureaucracies concerned, *Science and Technology and the Concept of Development. Utilization of the United Nations System in the Application of Science and Technology to Development.* Note by the Administrative Committee on Coordination (A/CONF.81/4/Add.1).

60. Ibid., pp. 46–83.

61. For a preconference attempt at outlining the relevant options for changing the institutional framework as part of a programme of action see Volker Rittberger, *Options for an Institutional Follow-up to the UNCSTD*, Science and Technology Working Papers Series No. 6 (New York: UNITAR, 1979), especially pp. 20–22.

62. On the position and function of the United Nations Director-General for Development and International Economic Cooperation see General Assembly Resolutions 32/197 of 20 December 1977 and 33/202 of 29 January 1979, in *GAOR, 32nd Session, Supplement No. 45*, pp. 121–127, and *GAOR, 33rd Session, Supplement No. 45*, pp. 129–131.

63. See Annex I, "Issues of the Draft Programme of Action on Which Agreement Was Not Reached at the Conference," and Annex II, "Texts Proposed by the Chairman of the First Committee on Some of the Issues in Annex I for Addition to the Vienna Programme of Action," in *Report*, op. cit. (note 1), pp. 84–97 and 98–101.

64. This gap has been described in Chapter 1, section 6.

65. Cf. Chapter 1, section 7.

66. The analysis of the process of consensus formation at UNCSTD, including its preparatory period, is based on participant observation by the author, who took part in the Conference process as a UNITAR representative. To the extent possible, the author's personal observations were checked against those of other participants.

67. This typology of strategies of programmatic consensus formation was first proposed by Ernst Haas, "International Integration: The European and the Universal Process," in *International Political Communities* (Garden City, N.Y.: Anchor Books, 1966), pp. 93–129, especially pp. 95–96.

68. The concept of policy legitimacy is used here in the sense given to it by Alexander L. George, "Domestic Constraints on Regime Change in U.S. Foreign Policy: The Need for Policy Legitimacy," in Ole R. Holsti et al. (eds.), *Change in the International System* (Boulder, Colo.: Westview Press, 1980), pp. 233–262, especially pp. 235–236.

69. *Report of the United Nations Conference on the Human Environment*, Stockholm, 5–16 June 1972 (New York: United Nations, 1973) (A/CONF.48/14/Rev.1; Sales No.: E.73.II.A.14).

70. *Report of the United Nations World Population Conference*, Bucharest, 19–30 August 1974 (New York: United Nations, 1975) (E/CONF.60/19; Sales No.: E.75.XIII.3).

71. *Report of the World Food Conference*, Rome, 5–16 November 1974 (New York: United Nations, 1975) (E/CONF.65/20; Sales No.: E.75.II.A.3). See also the following monographic analyses of these early United Nations conferences: Benno Engels, Khushi M. Khan, and Volker Matthies, *Weltwirtschaftsordnung am Wendepunkt: Konflikt oder Kooperation?* (Munich: Weltforum, 1975); Thomas G. Weiss and Robert S. Jordan, *The World Food Conference and Global Problem Solving* (New York: Praeger Publishers, 1976).

72. General Assembly Resolution 34/218 of 19 December 1979, in *GAOR, 34th Session, Supplement No. 46*, pp. 153–160.

73. Cf. paragraph 103 of the Vienna Programme in *Report of UNCSTD*, op. cit. (note 1), pp. 78–79.

74. Cf. General Assembly document A/34/587/Add. 1 of 23 November 1979.

75. General Assembly Resolution 34/218, section III, in *GAOR, 34th Session, Supplement No. 46*, p. 155.

76. This draft prospectus is contained in section IV of General Assembly document A/34/587 of 14 November 1979.

77. Cf. the draft resolution submitted by the delegation of India on behalf of the Group of 77, General Assembly document A/C.2/34/L.79/Add. 1 of 6 December 1979.

78. The Prospectus as agreed upon is found in the annex of General Assembly Resolution 34/218, in *GAOR, 34th Session, Supplement No. 46*, pp. 157–160.

79. See ibid., section V, p. 155. Earlier attempts at conducting such a study were made prior to UNCSTD; cf. *Cross-Organizational Analysis of Programme Activities Within the United Nations System Concerning Science and Technology for Development* (A/CONF.81/PC/CRP.13 and Add.1) and *Analysis of the Resources Available Within the United Nations System for Strengthening the Scientific and Technological Capacity of Developing Countries* (A/CONF.81/PC/44).

80. This point was stressed in a UNITAR seminar discussion on the restructuring of the economic and social sectors of the United Nations system. See *The Restructuring of the United Nations System.* Draft Report of a UNITAR Seminar, Schloss Hernstein, Austria, 12–15 July 1978, pp. 13–14. Cf. also John P. Renninger, "The Restructuring of the Economic and Social Sectors of the U.N. System: An Analysis." Unpublished paper, 1980, p. 26.

81. Following Renate Mayntz, "implementation" is used here in the sense of "the carrying out . . . of the programmes of action . . . which were elaborated in the process of policymaking. . . . The significance of implementation within the broader political process . . . lies in the banal, but equally incontestable fact that the political programmes determine only very incompletely the results of administrative action, i.e., their effects depend largely on the manner in which they are pursued." The analysis of implementation sets out to describe and explain the occurrence of discrepancies between programme objectives and the effects or results actually produced. In particular, such discrepancies may occur in the form of substantive implementation deficits, shifts in objectives, unintended selectivity or side effects of policy measures, and of simple, though consequential, time lags. See Renate Mayntz, "Die Implementation politischer Programme: Theoretische Überlegungen zu einem neuen Forschungsgebiet," *Die Verwaltung,* Vol. 10, No. 1 (1977), pp. 51–66; and Renate Mayntz (ed.), *Implementation politischer Programme* (Königstein: Athenäum-Hain-Scriptor-Hanstein, 1980).

82. See *Report of the Intergovernmental Committee on Science and Technology for Development,* First Session (28 January–1 February 1980), in *GAOR, 35th Session, Supplement No. 38,* pp. 1–15.

83. *Report of the Intergovernmental Committee on Science and Technology for Development,* Second Session (22 May–4 June 1980), in *GAOR, 35th Session, Supplement No. 37,* pp. 17–49.

84. On this and the following cf. *Progress Report on the Operations of the Interim Fund for Science and Technology for Development.* Note by the Administrator of the United Nations Programme of 8 May 1980 (A/CN.11/6).

85. This can be seen in the sections and paragraphs devoted to science and technology in the International Development Strategy for the Third United Nations Development Decade. Cf. General Assembly Decision S-11/23 of September 1980 in GAOR, 11th Special Session, Supplement No.3 (A/S-11/3), p. 7.

86. This is the title of a proposal submitted by the delegation of Belgium to the 11th Special Session of the United Nations General Assembly (A/S-11/AC.1/5 of 6 September 1980). The General Assembly at its 35th session has requested the Secretary-General and the competent organs of the United Nations to undertake a detailed study of this proposal.

Index